THE CHEMICAL SOCIETY
MONOGRAPHS FOR TEACHERS No. 9

Principles of Reaction Kinetics

(Revised Second Edition)

P. G. ASHMORE, MA, PhD
Professor of Physical Chemistry,
University of Manchester Institute of Science and Technology

LONDON: THE CHEMICAL SOCIETY

Monographs for Teachers

This is another publication in the series of Monographs for Teachers which was launched in 1959 by the Royal Institute of Chemistry. The initial aim of the series was to present concise and authoritative accounts of selected well-defined topics in chemistry for those who teach the subject at GCE Advanced level and above. This scope has now been widened to cover accounts of newer areas of chemistry or of interdisciplinary fields that make use of chemistry. Though intended primarily for teachers of chemistry, the monographs have proved of great value to a much wider readership, including students in further and higher education.

First published 1965; Second edition, 1967;
Reprinted with revisions, 1969; Revised second edition, 1973;
Reprinted March 1979.

Published by The Chemical Society, Burlington House, London W1V 0BN, and distributed by The Chemical Society, Distribution Centre, Blackhorse Road, Letchworth, Herts SG6 1HN

Printed by Butler & Tanner Ltd,
Frome and London

CONTENTS

1. Introductory Survey

The scientific study of reaction rates began over a hundred years ago. The earliest kinetic investigations were naturally made upon systems where the change of concentration of a reactant or product could easily be followed by chemical analysis over periods of many minutes or hours. A good deal of work continues on such slow reactions. On the one hand, it is important to know just how rapidly a product is being formed and to control the rate of formation, for example in some useful preparative reaction; on the other hand, a large body of quantitative knowledge about the rates and mechanisms of different types of reaction is an indispensable step in forming and testing theories of chemical reactivity. However, experimental techniques have so improved, particularly with the application of rapid physical methods of analysis, that investigations have now been extended to rapid reactions in solution and in the gas phase that are substantially completed in milliseconds or even microseconds. This has greatly improved our understanding of reactions as diverse as those between ions, enzyme-catalysed reactions proceeding in biological tissues, the reactions of free atoms and radicals formed by irradiating solutions or in the upper atmosphere, and high-temperature reactions in flames or shock waves.

The diversity of the reactions studied, and the variety of conditions of temperature, concentration and environment that are encountered, make it unlikely that there will be any single comprehensive theory of reaction kinetics comparable, for example, with the powerful thermodynamic treatment of chemical equilibria.

There are well-established equations of chemical thermodynamics that allow the calculation of the equilibrium constants of many reversible reactions from the properties of the molecules concerned, quite independently of direct experimental measurement of the constant. Thus, if the equilibrium involves relatively stable molecules, such as

$$CO + H_2O \rightleftharpoons CO_2 + H_2$$

the equilibrium constant, K_p, can be calculated from a knowledge of the standard free-energy change of the reaction, $\Delta G^\ominus$. This free-energy change can be determined from calorimetric measurements of the standard enthalpy change of the reaction, $\Delta H^\ominus$, and calorimetric determinations of the absolute entropy, $S^\ominus$, of each chemical at any required temperature, according to the equation

$$\Delta G^\ominus = \Delta H^\ominus - T\Delta S^\ominus = -RT \ln (K_p/\text{units})$$

$$\Delta S^\ominus = \sum_{\text{products}} S^\ominus - \sum_{\text{reactants}} S^\ominus$$

When the equilibrium involves free atoms or radicals, such as

$$Br_2 \rightleftharpoons 2Br$$

or

$$H + D_2 \rightleftharpoons HD + D$$

these standard free-energy changes cannot, of course, be determined calorimetrically. However, it is possible to evaluate the equilibrium constants for (atom + molecule) reactions in gases from equations developed in statistical mechanics, provided that quantities like the mass of each species, and the moment of inertia, fundamental vibration frequency and dissociation energy of each molecule are known. As these can be measured for diatomic molecules with great accuracy the calculated equilibrium constants of reactions like those shown above are often more accurate than the measured values. These calculations can be extended to more complex molecules or radicals with rather less accuracy, and to non-ideal systems by semi-empirical methods.

In contrast to this success, there is no basic theory that allows the prediction of the rates of chemical reactions with more than order-of-magnitude accuracy. The reasons for this can be seen by considering some simple reactions on a molecular scale, choosing reactions of atoms and molecules in the gas phase to avoid any complications due to the presence of solvent molecules.

Dissociation of simple molecules

Most chemical reactions involve the breaking of chemical bonds, the making of new bonds or both processes, although there are some exceptions like charge-transfer reactions. The simplest bond-breaking reaction is the dissociation of a diatomic molecule, like that of bromine, into free atoms:

$$Br_2 \rightarrow 2Br$$

In this reaction, energy ϵ_D is absorbed for each molecular step, or $D = L\epsilon_D$ for each mole of reaction (L is the Avogadro constant, $6.02 \times 10^{23} \text{mol}^{-1}$). This energy has to be put into each molecule as vibrational energy, increasing by quanta as shown schematically in *Fig.* 1*a*. In an assembly of molecules at equilibrium, in the absence of processes like dissociation, vibrational and translational energy are transferred by collisions; this exchange of energy maintains definite fractions of molecules with each allowed vibrational energy. If the total concentration of molecules is N per unit volume, with $N_0 \ldots N_\epsilon \ldots$ in levels $0 \ldots \epsilon \ldots$ respectively, there is a well-known expression first derived by Boltzmann relating N_ϵ to N_0:

$$N_\epsilon = N_0 \exp(-\epsilon/kT)$$

where k is Boltzmann's constant. Hence

$$N = \sum_\epsilon N_\epsilon = N_0 \sum_\epsilon \exp(-\epsilon/kT).$$

For molecules like bromine, where the vibrational energy-levels are spaced well apart, most molecules are in the lower levels and the summation does not differ greatly from unity. We can write $N = \beta N_0$, where β has been shown to increase slowly from 1.3 at room temperature to 2.9 at 1000 K, and so

$$N_\epsilon = \beta N \exp(-\epsilon/kT).$$

We may introduce the dissociation rather naïvely as a slight disturbance or perturbation of the equilibrium system, as depicted in *Fig.* 1*b*. (p. 6). We suppose that only molecules with the critical energy ϵ_D above the lowest energy can dissociate, and do so at a rate that scarcely affects the maintenance of the equilibrium population, N_{ϵ_D}, by the balance of activating collisions which raise the vibrational energy of less active molecules, and deactivating collisions which lower the energy of more active molecules.

The situation can be usefully compared to a barber's shop. The barber cuts hair at a fairly constant rate (corresponding to the actual dissociation) provided there are always customers waiting. The shop is kept full (corresponding to maintenance of the population N_{ϵ_D}) provided that the shop is in a busy street; many people arrive at the shop, only to turn away unless there is room to go in. It can easily be seen that the time each customer has to wait in the shop is equal to

$$\frac{\text{the number of customers waiting}}{\text{the number of haircuts per unit time}}$$

and hence the number of haircuts per unit time is

$$\frac{\text{the number of customers waiting}}{\text{time spent in waiting}}$$

In a similar manner, for the near-equilibrium conditions, the number of molecules dissociating per unit time is equal to

$$\frac{\text{population of energized state}}{\text{time spent in energized state}}$$

If each molecule with energy ϵ_D vibrates ν_D times per second and dissociates after one vibration, the time spent in the energized state is $1/\nu_D$, and hence

$$\text{Rate of dissociation} = \frac{N_{\epsilon_D}\ \text{(population)}}{1/\nu_D\ \text{(time)}}$$

$$= \nu_D\ \beta N \exp(-\epsilon_D/kT).$$

Here is a relatively simple theoretical expression for the rate of dissociation, in numbers of molecules dissociated per unit volume per second. Notice that the right-hand side of the equation consists of four factors: a concentration term, N, which in a constant volume would be proportional to the pressure of bromine gas and independent of the temperature, a frequency ν_D, which introduces the time factor, a dimensionless factor β that increases very slowly with increase in temperature and a dimensionless exponential factor which indicates the fraction of molecules that have enough energy to dissociate into atoms. Each of the factors can be evaluated, and so the form and the magnitude of the predicted rate expression can be tested against experimental results.

However, it has not been possible until recently to obtain experimental results. At low temperatures the extent of dissociation is so small that it cannot be measured. At high temperatures the rate of dissociation is so rapid that the system reaches equilibrium in the time taken to raise the temperature by conventional methods. The advent of the chemical shock-tube, in which a gas can be heated extremely rapidly and the subsequent reactions followed, has altered the position, and the rates of dissociation of gases like Br_2 and O_2, alone or mixed with inert gases like argon or nitrogen, have been measured at various temperatures between 1000 K and 2500 K (*see, e.g.*, S. H. Bauer, *Science*, 1963, **141**, 867).

Perhaps the most significant result is that the rate does depend exponentially upon the temperature as the equation predicts. The practical effect of this exponential form of the temperature factor is that the rate changes very rapidly with the temperature. For bromine, ϵ_D/k is 22 800 K, being equal to D/R where R is the gas constant (8.31 J K^{-1} mol^{-1}) and D is the dissociation energy (189 kJ mol^{-1}). Thus the exponential factor takes the following approximate values:

Temperature, K	300	1000	5000
Factor	10^{-33}	10^{-10}	10^{-2}

The rate of dissociation increases by a factor of 10^{23} between room temperature and 1000 K. Notice that this factor overwhelms the comparatively trivial increase in the sum, β.

The simple expression fails, however, when the other factors are examined. The experiments show that in mixtures of bromine with an excess of an inert gas like argon the rate is proportional to the concentration of bromine gas, but it is also proportional to the concentration of the inert gas. In bromine alone, the rate is proportional to the square of the bromine concentration. In either case, the extra concentration factor means that the remaining factor, ν_D, has to be considerably modified to keep the dimensions of the equation correct.

Let us return for a moment to the analogy of the barber's shop. We assumed that the waiting chairs were always filled because the supply of potential customers was more than adequate to keep pace with the barber's output. Suppose, however, that the street outside is nearly empty. The number of haircuts per unit time ceases to depend upon events within the shop and depends instead upon the infrequent arrival of suitable customers at the door. Correspondingly, if there are few bromine molecules present in a vessel the population of energized molecules is severely depleted by the actual dissociation, and the whole rate becomes dependent upon the rate of energization. This is a collision process, and in bromine alone the rate of collision depends upon the square of the bromine concentration. With an excess of argon, collisions between argon and bromine molecules are the most frequent, and the rate of energization depends on the product of the two concentrations. The complete process can be represented by

$$Br_2 + M \rightleftharpoons Br_2^* + M$$

$$Br_2^* \rightarrow 2Br$$

where M represents any species present and Br_2^* an activated molecule of bromine. We shall see in a later section (p. 54) that as the total pressure is increased, either by adding more bromine or more argon, the rate should cease to depend upon the product of two concentrations and should be proportional to a single concentration. The transition corresponds, in our analogy, to the street getting busier and the waiting chairs filling up, until the rate of cutting again becomes the limiting process. With simple molecules like bromine, it has not been possible to demonstrate this transition. With more complex molecules, however, the transition occurs within ranges of pressure that are experimentally accessible. The larger molecules can collect energy in many vibrations and redistribute it, so that the chance of the necessary energy reaching the bond which has to be broken is greater in a more complex molecule than in a smaller molecule with the same rate of collision.

Two very important principles emerge from this brief consideration of the rate of a dissociation which looks extremely simple from the chemical equation. The first is the concept of energization or activation before even simple reactions occur; as will be explained later at greater length, the idea is applicable to nearly all types of reaction. The second is that apparently simple reactions can consist of several stages, and that different stages may control the rate under different conditions.

Recombination of free atoms

Consider a 'head-on' collision between two chemically inert argon atoms. At distances of several atomic diameters, there are weak attractive forces between the atoms. These forces slightly increase the kinetic energy of approach, with a corresponding loss of potential energy shown by the dip in the curve in *Fig.* 2*a*. As the atoms approach each other more closely, strong repulsions occur between the electrons. In terms of the Pauli principle, the electrons of atom 1 cannot enter the filled orbitals of atom 2, and vice versa, without the expenditure of considerable amounts of energy. The potential energy of the system rises steeply, and the kinetic energy falls until the relative motion stops at the distance of closest approach, σ, equal to the collision diameter. The atoms then accelerate away from each other and an 'elastic' collision has occurred.

If, however, two free bromine atoms approach each other, the events are different at very close approaches. There are atomic orbitals that are not fully occupied, and a chemical bond can be formed by the reverse of the dissociation process. Such a collision is 'inelastic'. The potential-energy changes follow a composite of *Figs.* 1*a* and 2*a*. The most likely pattern is shown in *Fig.* 2*b*, which assumes that the two atoms formed by dissociation in 1*a*

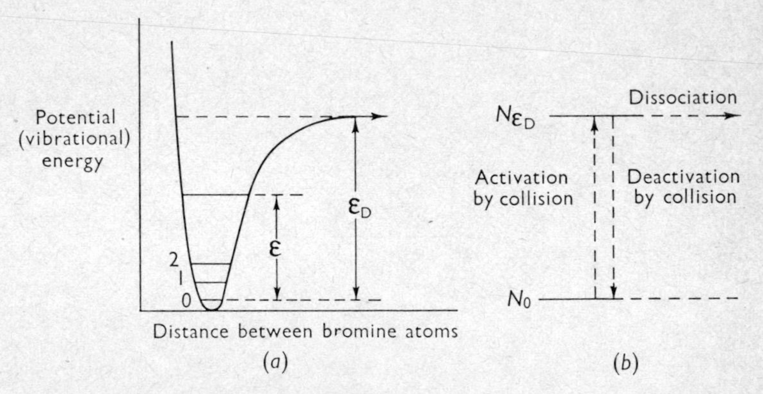

FIG. 1. Schematic representation of (*a*) the vibrational energy of bromine molecules; (*b*) the population of levels

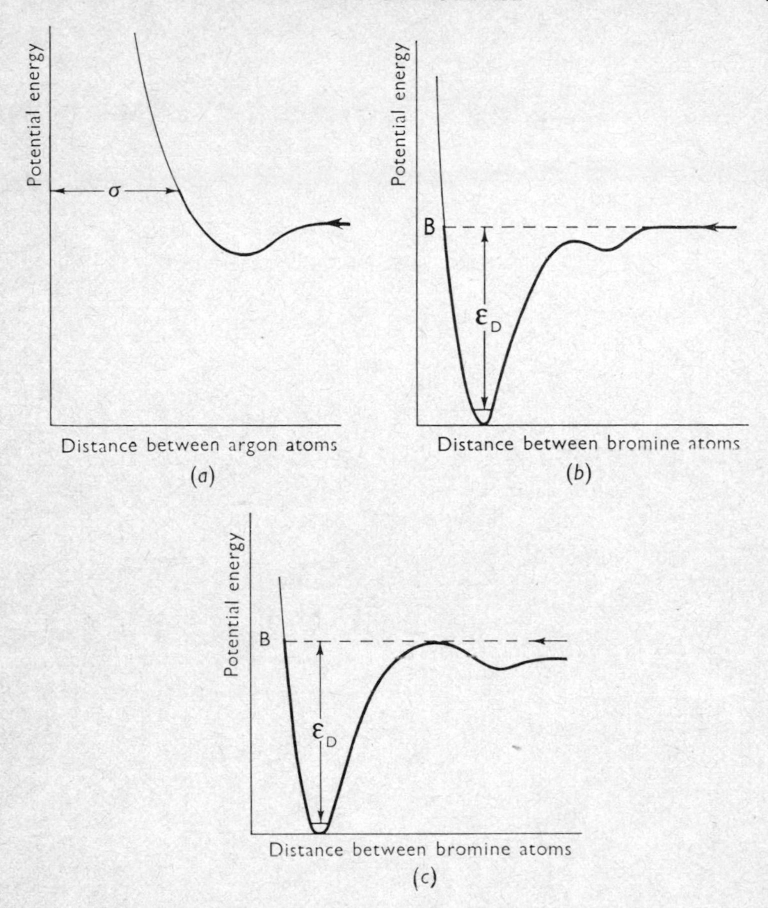

FIG. 2. Schematic representations of potential-energy changes in (*a*) Elastic collision of argon atoms; (*b*) Inelastic collision—likely pattern for bromine atoms; (*c*) Inelastic collision—unlikely pattern for bromine atoms

are, like the two approaching in *Fig. 2a*, of average kinetic energy. A less likely pattern is shown in *Fig. 2c*, demanding an *energy barrier* for recombination which can only be surmounted by *activated* atoms with extra kinetic energy. Recently, experiments have shown that no activation energy is required for the process, so *Fig. 2b* is correct. In outline, these 'flash photolysis' experiments* consist of irradiating bromine molecules with a sharp pulse of light which dissociates the molecules into free atoms. The formation of molecules is then followed by measuring the absorption

* *See e.g.*, R. G. W. Norrish and B. A. Thrush, *Q. Rev. chem. Soc.*, 1956, **10**, 149.

of light of longer wavelength which does not re-dissociate the molecules. From these measurements the variation of the molecular concentration with time can be calculated, the atomic concentration being calculated for each time from the mass balance or the atom balance.

If no activation energy were needed, the rate of formation of bromine molecules would equal the number of collisions between bromine atoms per unit volume per second. This can be shown, from the kinetic theory of gases, to be equal to

$$2\sigma_{Br}^2 N_{Br}^2 \sqrt{\frac{\pi RT}{M_{Br}}}$$

where σ_{Br} is the collision diameter, M_{Br} is the molar mass of atomic bromine, R is the universal gas constant, and N_{Br} is the number of bromine atoms per unit volume.

Unfortunately, experiments show that although the rate of formation of bromine molecules is proportional to N_{Br}^2, it is also proportional to the concentration of any other gas molecules present in excess, for example added argon or molecular bromine. Moreover, although the collision formula predicts that the rate will slowly increase as the temperature is raised, other things being equal, experiments quite definitely show that the rate slowly *decreases* as the temperature is raised.

These results have been interpreted as follows. If two bromine atoms meet and combine, their potential energy changes along the curve in *Fig.* 2*b* from right to left. The incipient molecule at B, however, will fly apart again on the next vibration, unless it can lose some energy in the meantime. It can only do this by collision, so it is stabilized by collision with the species (molecular bromine or argon, for example) in excess, and hence the rate depends on a third concentration factor.

The theoretical problem thus becomes the calculation of triple collisions of the type

$$Br + Br + M \rightarrow Br_2 + M$$

and it has proved very difficult to do this accurately. Another approach is to consider that the reaction proceeds in two stages. The first is reversible and forms an unstable species like Br_2^*, which either dissociates or is deactivated in a collision with M. However, any calculations on this model are very difficult since little is known about the concentration of the unstable species or about the efficiency of the energy exchange:

$$Br + Br \rightleftharpoons Br_2^*$$
$$Br_2^* + M \rightarrow Br_2 + M$$

Thus for both dissociation and recombination reactions it is possible to suggest molecular models for the processes that are satisfactory in some ways, but not possible to carry out any accurate theoretical calculations of the rates to compare with the experimental values.

Perhaps more important than this limitation, however, is the implication that some of the simplest possible reactions from a chemical viewpoint do not proceed as the chemical equation suggests at first sight. More than one molecular step is involved in the dissociation, and more than one may be involved in the recombination.

It turns out that many reactions in gases and in solution, and heterogeneous reactions that proceed at the interfaces of different phases, are *complex* and proceed by stages. Each stage is an *elementary* reaction in which the number of bonds broken and made is small. In the early stages the reactants form intermediates which may be other molecules, ions, free radicals or atoms. These intermediates react in later stages to yield the products. Thus in most thermal decompositions free radicals are formed by dissociation of a molecule, and these decompose or react with the parent molecules to generate the final products. All catalysed reactions are complex, involving reaction between reactant and catalyst; thus acid catalysis involves donation of a proton to a reactant molecule to form an intermediate ion, which by further reactions gives the products and re-forms the acid catalyst. In heterogeneous reactions surface compounds are formed in which fragments of the reactant are bonded to the surface, to undergo further reaction to products.

This distinction between elementary reactions that proceed in a single step and complex reactions that proceed by a number of elementary steps is fundamental in all modern approaches to reaction kinetics.

It will be convenient to consider briefly in the rest of this chapter the theory of elementary, single-stage reactions. This theory attempts to relate the rate of such reactions to the concentration of the reactants, the temperature, and the chemical properties and molecular structure of the reactants. Later we shall discuss the types of pattern observed in complex reactions, the connexions between the rate of the overall reaction and the rates of the constituent steps, the nature and structure of the intermediates and the reasons why particular reactions follow one pattern and not another.

A simple elementary reaction

We shall now consider a typical elementary reaction, an exchange or transfer reaction in which one bond is broken and one new bond is made

$$H + D_2 \rightarrow HD + D \quad \text{.. .. ..} \quad (R1)$$

The rate of this reaction can conveniently be studied at temperatures around 1000 K, where the dissociation of molecular hydrogen produces a low, steady concentration of hydrogen atoms. These react slowly in reaction R1, and the extent of the exchange is measured by the change in the thermal conductivity of the gas mixture. The detailed treatment is complicated by the occurrence of the reaction

$$D + H_2 \rightarrow HD + H \qquad .. \qquad .. \quad (R1')$$

and by the reversible nature of R1 and R1′, but it has been possible to measure the rates of all reactions of this type. The rates have also been measured at room temperature by generating hydrogen atoms photochemically or with an electric discharge.

On the molecular scale the reaction might occur in a variety of ways. At one extreme the deuterium molecule could be dissociated, followed by recombination of the hydrogen atom with one deuterium atom. This dissociation requires the expenditure of a large amount of energy, ϵ_D, relative to the average kinetic energy of the particles at normal temperatures—about 100 times at room temperature and about 17 times at 2000 K. A similar quantity of energy would be released by the recombination, but if the processes were separate this energy could not provide or reduce the original expenditure (*Fig.* 3*a*). However, if the H–D bond is formed at the same time as the D–D bond is ruptured, much less energy is required. The amount required can be shown to depend on the spatial distribution of the nuclei during reaction. As the hydrogen atom approaches the deuterium molecule along some arbitrary direction shown by the dotted line in *Fig.* 3*b*, the potential energy of the system falls a little, as in *Fig.* 2*a*; it then increases because the deuterium molecular orbital of least energy is fully occupied, and there is strong repulsion between the electrons in this orbital and the electron in the hydrogen atomic orbital. At some potential energy, ϵ_α, the electron shift occurs which breaks the D–D bond and forms the H–D bond; the potential energy then falls as the HD molecule and the D atom separate. The energy ϵ_α represents an *energy barrier* to the reaction which can only be overcome by expenditure of kinetic energy. The energy that is effective is the kinetic energy of motion along the line of approach. Only those pairs of particles with kinetic energy in excess of ϵ_α can react when colliding along that line of approach; the collisions with less than ϵ_α are 'elastic' like those of the argon atoms previously considered, and the particles then bounce apart. The fraction of collisions where this kinetic energy is in excess of ϵ_α is found to be $\exp(-\epsilon_\alpha/kT)$ where k, as before, is Boltzmann's constant; this fraction diminishes rapidly as ϵ_α increases. Now the height of the barrier varies with α, and is

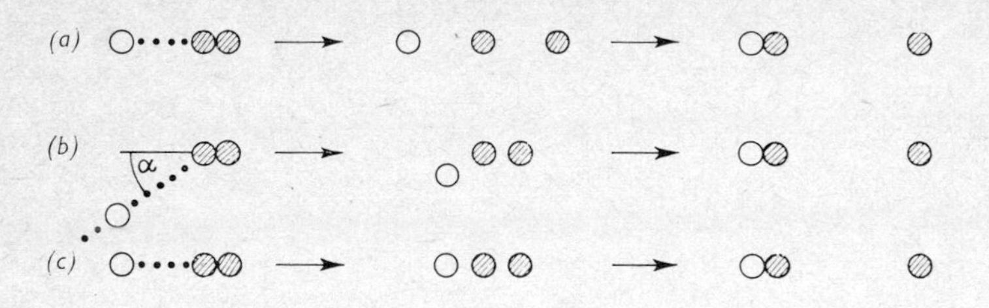

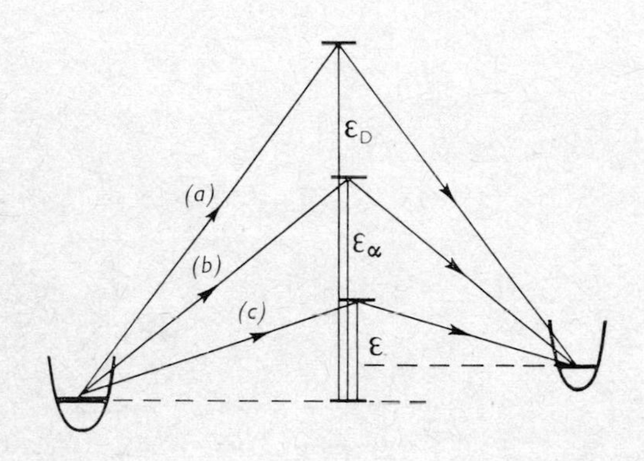

FIG. 3. The configurations and relative potential energies for reaction paths (*a*) where D_2 dissociates before HD is formed; (*b*) HD is formed as D_2 dissociates, with H approaching D_2 at angle α; (*c*) as for (*b*) with $\alpha = 0$. H ○ D ◍

found to be a minimum, ϵ, for this 'three-centre' reaction when the H atom approaches along the D–D axis ($\alpha = 0$). Consequently this path contributes far more to the reaction than other possible paths with higher ϵ_α, because of the exponential form of the factor, and the path assumes a special importance for the reaction. The particular configuration of atoms corresponding to the lowest barrier (*see Fig.* 3*c*) is called the *transition state* or *transition complex* of the reaction. To reach this configuration the reactants must be 'activated' and so share between them kinetic energy of at least ϵ for each transition complex or $E = L\epsilon$ for each mole of transition complex. (L is the Avogadro constant, 6.02×10^{23} mol^{-1}.) To develop this model of the molecular reaction further, it is necessary to calculate the fraction of collisions with suitably orientated particles and a total kinetic energy greater than ϵ, or alternatively to calculate the concentration of transition complexes and the rate at which they form product molecules.

The first of these approaches was introduced in 1918 by W. C. McC. Lewis as the collision theory of reaction rates. The total number of collisions, Z, in one second in each unit volume can be calculated from the kinetic theory of gases by a formula similar to that given for collisions of bromine atoms (p. 8); for reaction R1 it would be proportional to $[H][D_2]$. Hence $Z = Z'[H][D_2]$; it may be noted here that Z' does not change very much with the type of molecule concerned, because increase in size leads to a bigger cross-section for collision but a smaller velocity of approach. The fraction of collisions with kinetic energy along the line of centres in excess of ϵ is, as mentioned, $\exp(-\epsilon/kT)$. The need for correct orientation of the molecules ($\alpha = 0$) can be allowed for in a steric factor, P. The theory thus predicts that the rate of reaction R1 would be

$$\text{rate} = PZ'\,[H][D_2]\exp(-\epsilon/kT) \quad .. \quad .. \quad (1)$$

The theory was tested by Hinshelwood and others for many simple gas reactions between two molecules, such as

$$H_2 + I_2 \rightarrow 2HI \quad .. \quad .. \quad .. \quad (R2)$$

$$2NO_2 \rightarrow 2NO + O_2$$

$$2HI \rightarrow H_2 + I_2 \quad .. \quad .. \quad .. \quad (R3$$

We shall not examine the tests in detail, but it is important to note that no attempt was made in the tests to calculate ϵ from first principles. Equation 1 shows that ϵ can, in principle, be calculated in two ways for reaction R3:

(*a*) if the rates are measured for chosen concentrations of HI at a fixed temperature, and the pre-exponential factor calculated from the kinetic theory of gases and the assumption $P \sim 1$ for these simple reactions, the exponential factor and hence ϵ can be determined;

(*b*) if rates are measured at a number of temperatures, keeping [HI] constant and noting that Z' is proportional to $T^{1/2}$, eqn 1 shows that

$$\ln(\text{rate}_1 T_2/\,\text{rate}_2 T_1) = -\,\epsilon/kT_1 + \epsilon/kT_2$$

Thus, a plot of the l.h.s. against $1/T_2$, with T_2 and rate_2 fixed, should be linear, and the slope should be $-(\epsilon/k)$. For each of these simple reactions the two calculations of ϵ agreed remarkably well. The theory was extended by Kassel and Lindemann to reactions of more complicated molecules, and by Moelwyn-Hughes to reactions between solutes in dilute solutions. These extensions were less successful, however. If ϵ was determined from method (*b*), the calculation in (*a*) required very marked changes in P from reaction to reaction. Certain trends were noted relating changes in P to the complexity of the reacting molecules, but there seemed to be no independent way of predicting these changes. This limitation, and the need to determine ϵ from rate measurements, means that rates cannot be predicted by the collision theory from first principles.

The second approach really developed from the concept of activation introduced long ago by Arrhenius. He postulated that chemical reaction occurs after normal molecules are changed to activated molecules; these activated molecules usually revert to normal molecules and are considered to be in equilibrium with them, but occasionally an activated molecule changes to the products of reaction. Thus for a reaction A + B → products, the model is

$$\mathrm{A} + \mathrm{B} \rightleftharpoons \text{activated molecules} \rightarrow \text{products}$$

and, from the equilibrium relationship for gases or dilute solutions,

$$\frac{[\text{activated molecules}]}{[\mathrm{A}][\mathrm{B}]} = K_c$$

If the rate of the final step is proportional to [activated molecules]

$$\text{rate} = \alpha K_c[\mathrm{A}][\mathrm{B}] \quad .. \quad .. \quad .. \quad (2)$$

This equation predicts that the rate is proportional to the product of the concentrations of all the reactants, *i.e.*

$$\text{rate} = k \times \prod_{\text{reactants}} (\text{concn})$$

where k is the specific reaction rate (the value of the rate when each concentration is made unity) or *rate constant* at constant temperature. This is in keeping with the older empirical law of Guldberg and Waage, but the new formulation allows the prediction of the effect of temperature changes. If $k = \alpha K_c$ and α is independent of temperature

$$\ln(k/\text{units}) = \ln(\alpha/\text{units}) + \ln(K_c/\text{units})$$

$$\therefore \frac{\mathrm{d}\ln k}{\mathrm{d}T} = \frac{\mathrm{d}\ln K_c}{\mathrm{d}T} = \frac{\Delta U_c^{\ominus}}{RT^2}$$

where $\Delta U_c^{\ominus}$ is the change in the standard internal energy in forming activated molecules from reactants. If $\Delta U_c^{\ominus}$ is independent of temperature, which is a reasonable assumption, integration of the equation gives

$$\ln(k/\text{units}) = \text{constant} - \frac{\Delta U_c^{\ominus}}{RT}$$

Arrhenius examined measurements of the rates of several reactions in solution and showed that the experimental results fitted an equation

$$\ln (k/\text{units}) = \text{constant} - \frac{E}{RT} \quad \ldots \quad (3)$$

where E is the *experimental activation energy*, representing the value of $\Delta U_c^\ominus$ for the model, characteristic of each reaction.

Unfortunately no method could be found for evaluating $\Delta U_c^\ominus$ or the constant in eqn 3 independently of the rate measurements, so this form of the activation theory does not allow the prediction of rates from the properties of the molecules of reactants.

A considerable modification of the theory of activation was made in the nineteen-thirties by Wigner, Polanyi and Eyring who identified 'activated molecules' with 'transition complexes' (p. 11) and showed how to calculate the equilibrium constant from the equations of statistical mechanics mentioned earlier. The same factor $\exp(-\epsilon/kT)$ appears in this calculation (strictly the energy, ϵ_0, of formation of the complex if the equilibrium could exist at absolute zero of temperature) and the rate is again proportional to the concentrations of the reacting species, *i.e.* to $[H][D_2]$ for reaction R1. The frequency of passage of complex to products is set equal to the frequency of vibration of the bond about to break. This is usually 10^{12} or 10^{13} s^{-1} and so the lifetime of the complex is only 10^{-13} to 10^{-12} s. The properties of the reactants and the transition complex that appear in the calculation of the equilibrium constant, apart from ϵ, are the moments of inertia, I, the vibration frequencies, ν, and the molecular masses, m. These properties and the temperature occur in the final expression for the rate, which for R1 is

$$\text{rate} = [H][D_2] \frac{kT}{h} \, f(m,I,\nu,T) \exp(-\epsilon_0/kT)$$

(h is Planck's constant). This equation is somewhat like eqn 1, but differs from it in that the function $f(m,I,\nu,T)$ can be calculated if the properties of the transition complex can be determined*. Moreover, it allows a considerable variation in the value of the function as the complexities of the reacting molecules are changed. Without pursuing this point in detail, it can be said that these variations are closely in line with the observed changes in rates. Thus the theory is more satisfactory than the other approaches. In principle, it allows the calculation of rates from the properties of the reactants and the transition complex. In practice, however,

---

* The function takes somewhat different forms which depend upon the complexity of the reactants. *See* Laidler, listed in Introductory Reading, p. 79.

it is difficult to determine the properties of the transition complex with sufficient accuracy to be really useful. The complex has too short a life to be observed experimentally. The moments of inertia have to be estimated by calculation for likely structures of the complex, and the frequencies of vibration estimated by analogy with the frequencies found for stable molecules of similar structure. Another major stumbling block is the accurate estimation of ϵ_0, which appears in the exponential factor and so greatly affects the rate. Again, in principle, ϵ_0 could be calculated for reactions like R1 by solving the Schrödinger wave-equation for the system of three nuclei and three electrons for all relevant positions of the nuclei, so mapping out the energy profiles in order to find the minimum height of the barrier. The calculations are prohibitively long even with modern computing aids, and would be even more difficult for more complex reactants. Several approximations have been used by Polanyi, Eyring and later investigators, but they all introduce arbitrary parameters and the predicted rates of reaction often differ from the rates observed experimentally by several orders of magnitude.

The problems are even more formidable for reactions in solution, particularly in polar ionizing solvents. The substitution reaction

$$OH^- + CH_3Br \rightarrow CH_3OH + Br \quad .. \quad .. \quad (R4)$$

is thought to proceed, like R1, through a single transition complex. The potential-energy change during the formation of the complex does not only involve the reactants, however; in molecular detail the structure of the solvent around the reactant particles will change as they form the complex. It is not possible to predict quantitatively the effect of this change, thus putting farther out of reach any accurate theoretical prediction of the rate of R4 and reactions like it.

It will be observed that the collision theory and the transition-state theory predict that for a simple reaction the rate law is of the form

$$\text{rate} = A' \times \prod_{\text{reactants}} (\text{concn}) \times \exp(-\epsilon/kT)$$

but predict slightly different forms for the coefficient A'. However, when examined in detail they agree that the effect of temperature on A' is slight compared with the effect on the exponential term unless ϵ is very small.

Now the experimental evidence collected by Arrhenius, and a vast amount of later work, shows that for simple single-stage reactions the rate law is of the form

$$\text{rate} = k \times \prod_{\text{reactants}} (\text{concn}) \quad .. \quad .. \quad .. \quad (4)$$

where k is dependent on temperature according to eqn 3; this can be rewritten as

$$k = A \exp(-E/RT) \quad \ldots \quad \ldots \quad \ldots \quad \ldots \quad (5)$$

where A is approximately independent of temperature. Thus with some reservations A can be identified with A' and E/R with ϵ/k. The reservations are necessary because when examined in more detail the theories suggest different temperature-dependencies for A' and different meanings for ϵ. Unfortunately experimental techniques are not sufficiently refined to discriminate between the different predictions, and the theories are not sufficiently accurate to predict ϵ. Consequently it is generally agreed that eqn 5 should be taken to represent the most convenient and reasonably reliable expression for showing the dependence of rate constants and the rates of reaction on temperature. It is a concise representation of the theory of activation, and the quantities A and E are used to characterize the rate constant of each elementary reaction, as described in Chapter 2.

Elementary and complex reactions

The need to distinguish between elementary reactions that proceed in a single step and complex reactions that proceed by a number of elementary steps has already been emphasized. The concept of a transition complex for each elementary step can help us to decide whether a reaction is likely to proceed in one step or to need several stages. It is generally found that a single step is only feasible if it is possible to construct for it a transition complex in which the atomic spacings and positions are only slightly changed from those in the reactant molecules. For example, consider two reactions of aliphatic ethers, both appearing to be complex changes:

$$CH_2{=}CH{\cdot}O{\cdot}CH_2{\cdot}CH{=}CH_2 \rightarrow CH_2{=}CH{\cdot}CH_2CH_2{\cdot}CHO$$

$$CH_3{\cdot}O{\cdot}CH_3 \rightarrow CH_4 + CO + H_2$$

In the rearrangement of vinyl allyl ether to 3-vinyl propionaldehyde it is possible to construct a reasonable cyclic transition complex:

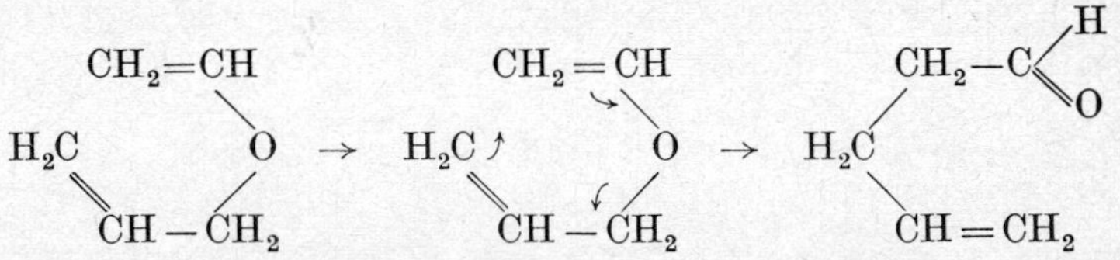

```
                                                              H
  CH2=CH              CH2=CH                    CH2-C<
        \               ↘\                     /      \\
H2C      O   →   H2C ↗     O   →   H2C              O
  \\    /          \\   ↙ /            \
   CH - CH2         CH - CH2           CH=CH2
```

whereas with the reaction proposed for the dimethyl ether it is impossible to construct a transition complex unless two of the hydrogen atoms on one methyl radical form the molecular hydrogen, the third being transferred to the other methyl to form methane. It is very unlikely that hydrogen elimination occurs like this. Consequently, the first reaction can proceed in one stage, but it is unlikely that the second does. There is good experimental evidence that the actual reactions follow the predicted patterns, the first being a simple molecular rearrangement proceeding readily at 200 °C, the second being a chain reaction which involves several steps before the products appear. The initiation step of the chain reaction is probably

$$CH_3{\cdot}O{\cdot}CH_3 \xrightarrow{1} CH_3{\cdot}O + CH_3$$

and this is followed by the propagation steps

$$CH_3 + CH_3{\cdot}O{\cdot}CH_3 \xrightarrow{2} CH_4 + CH_3{\cdot}O{\cdot}CH_2$$

$$CH_3{\cdot}O{\cdot}CH_2 \xrightarrow{3} CH_3 + CH_2O$$

with further reactions of the CH_2O to give CO and H_2 which, with CH_4, form the products. It can be seen that the chain centre CH_3 is consumed in the first propagating step but regenerated in the second, so that the sequence of propagating reactions can be repeated many times for each initiating reaction 1. However, the reaction does not 'explode' because the chain centres are removed in termination steps such as

$$CH_3 + CH_3{\cdot}O \rightarrow CH_3{\cdot}O{\cdot}CH_3$$

Because the initiation reaction involves breaking a strong C–O bond with no assistance from further bond-formation it has a high energy barrier, ϵ, so that the factor $\exp(-\epsilon/kT)$ is very small except at high temperatures; in agreement with this, the decomposition is very slow unless the temperature is above 500 °C.

We have here examples of two extreme patterns of reaction. The first proceeds through a single-stage elementary reaction with a well-defined transition complex. The second proceeds by repeating a succession of reaction steps, each in itself an elementary reaction with its own transition complex.

It is most essential to distinguish between intermediates in a reaction and transition complexes. The lifetime of the transition complex is only 10^{-13} to 10^{-12} s, and there is very little prospect of measuring the concentration of the transition complexes in an elementary reaction.

By an intermediate we mean a species that has a lifetime many magnitudes greater than 10^{-12} s. Each intermediate is formed through a transition complex and reacts with other species by forming another transition complex.

As mentioned earlier, it is not possible to predict accurately the rates of elementary reactions from first principles. Consequently, although the rates of complex reactions can usually be related to the rates of the individual steps, it is impossible to predict accurately the overall rate. Our knowledge of rates and mechanisms has therefore been built up in a semi-empirical way, using the theoretical concepts to guide the experimental search for specific information about each reaction. The experimental investigations of rates, and the effects of changes of temperature, concentrations and reaction-vessel surface, provide the basic information about the pattern of reaction. The information leads to *rate equations* or *rate laws* (eqn 1 is a simple example) connecting the rates of reaction with the concentrations of reactants, catalysts and products and with the temperature. As will be explained later, certain forms of the rate laws are indicative of certain patterns of reaction. A simple rate law like (1) is usually indicative of an elementary reaction. Complicated rate laws, like those given on pages 51 and 72, are always indicative of complex mechanisms. In some conditions a complex rate law can reduce to a simple equation and the complex reaction then masquerades as an elementary one.* From the detailed study of complex reactions it is sometimes possible to derive the rates of the elementary reactions in the proposed mechanism, even when these involve free atoms, radicals or ions. A useful body of information about elementary reactions involving unstable species has thus been added to our knowledge of the limited number of elementary reactions involving stable molecules. The extension and examination of this information in different ways will, it is hoped, indicate how fundamental theories of elementary reactions can best be improved.

---

* For many years reaction R2 has been regarded as a single-stage elementary reaction proceeding through a four-centre transition complex, the reverse reaction R3 proceeding through the same complex:

$$H_2 + I_2 \rightleftharpoons \begin{matrix} H & \text{- -} & I \\ \vdots & & \vdots \\ H & \text{- -} & I \end{matrix} \rightleftharpoons 2HI \qquad \text{(R2, R3)}$$

J. H. Sullivan has recently studied the photochemical reaction at lower temperature (*J. chem. Phys.*, 1967, **46**, 73) and has concluded that the thermal reaction could proceed almost entirely through an equilibrated dissociation $I_2 \rightleftharpoons 2I$ followed by a rate-determining step R2′:

$$H_2 + 2I \rightarrow 2HI \qquad (\text{rate:} -d[H_2]/dt = k'[H_2][I]^2) \quad (\text{R2}')$$

Sullivan has shown that the mechanism predicts an overall rate constant equal to Kk_2', where K is the equilibrium constant for the dissociation, and that the experimental rate constant of the thermal reaction is equal to Kk_2' at several temperatures.

2. Rates and Rate Laws for Elementary Reactions

In this chapter we shall indicate in more detail how the rates of elementary reactions are expressed, and give some examples of the ways in which the rates of some typical elementary reactions have been found to depend upon the reactant concentrations and the temperature. The discussion of general methods of measuring the rates and the treatment of the experimental data are postponed until an explanation has been given of the integration of typical rate equations.

Methods of expressing the rates

In order to be able to deal with reactions taking place in either constant volume systems or in flow systems in which concentrations may change because of pressure changes as well as by reaction, it is best to define rate of reaction per unit volume as $-(1/V)\mathrm{d}n_\mathrm{R}/\mathrm{d}t$ or $(1/V)\mathrm{d}n_\mathrm{P}/\mathrm{d}t$ where n_R, n_P are the number of moles of reactant R or product P in volume V.

In the past, the preponderance of systems operated at effectively constant volume led to the rate of reaction being automatically referred to unit volume and so defined as $-\mathrm{d}[\mathrm{R}]/\mathrm{d}t$ or $\mathrm{d}[\mathrm{P}]/\mathrm{d}t$ where $[\mathrm{R}] = n_\mathrm{R}/V$ and $[\mathrm{P}] = n_\mathrm{P}/V$.

If this last definition of rate is adopted, there is no ambiguity for reactions like R4 (p. 15) because the rate is equally expressed as $-\mathrm{d}[\mathrm{OH}^-]/\mathrm{d}t = -\mathrm{d}[\mathrm{CH_3Br}]/\mathrm{d}t = \mathrm{d}[\mathrm{CH_3OH}]/\mathrm{d}t = \mathrm{d}[\mathrm{Br}^-]/\mathrm{d}t$.

In the forward step of reaction R2, however, two molecules of hydrogen iodide are produced for every molecule of hydrogen or iodine disappearing, so that

$$\mathrm{d}[\mathrm{HI}]/\mathrm{d}t = 2(-\mathrm{d}[\mathrm{H_2}]/\mathrm{d}t) = 2(-\mathrm{d}[\mathrm{I_2}]/\mathrm{d}t)$$

Similar problems arise in all reactions where the stoichiometric numbers differ from unity, as in the general reaction

$$a\mathrm{A} + b\mathrm{B} \rightarrow c\mathrm{C} + d\mathrm{D}$$

An old proposal to avoid these ambiguities is currently finding much favour, namely to define the rate of reaction per unit volume as

$$-\frac{1}{aV}\frac{\mathrm{d}n_\mathrm{A}}{\mathrm{d}t} = -\frac{1}{bV}\frac{\mathrm{d}n_\mathrm{B}}{\mathrm{d}t} = \frac{1}{cV}\frac{\mathrm{d}n_\mathrm{C}}{\mathrm{d}t} = \frac{1}{dV}\frac{\mathrm{d}n_\mathrm{D}}{\mathrm{d}t}$$

or for constant volume conditions as

$$-\frac{1}{a}\frac{\mathrm{d}[\mathrm{A}]}{\mathrm{d}t} = -\frac{1}{b}\frac{\mathrm{d}[\mathrm{B}]}{\mathrm{d}t} = \frac{1}{c}\frac{\mathrm{d}[\mathrm{C}]}{\mathrm{d}t} = \frac{1}{d}\frac{\mathrm{d}[\mathrm{D}]}{\mathrm{d}t}$$

Thus for the ammonia synthesis $N_2(g) + 3H_2(g) \rightarrow 2NH_3(g)$ the forward rate at constant volume would be defined as

$$-\frac{\mathrm{d}[\mathrm{N_2}]}{\mathrm{d}t} = -\frac{1}{3}\frac{\mathrm{d}[\mathrm{H_2}]}{\mathrm{d}t} = \frac{1}{2}\frac{\mathrm{d}[\mathrm{NH_3}]}{\mathrm{d}t}$$

It should be remarked that the almost universal practice adopted by chemists of quoting 'rates of reaction per unit volume' as simply 'rates of reaction' has been criticized and alternative definitions proposed.*

Units for measuring the rates

The numerical value of the rate will depend upon the units used for concentration and for time. Although seconds, minutes and even hours have been used in the past, there is now universal agreement that the second is the preferred unit of time. For concentration, investigators working on the kinetics of reactions in solution seem to prefer moles per litre (mol $\mathrm{l^{-1}}$) whereas those working on gas reactions often use moles per cubic centimetre (mol $\mathrm{cm^{-3}}$) and also molecules per cubic centimetre (molecules $\mathrm{cm^{-3}}$) for ease of comparison with kinetic-theory expressions. The trend may well be towards the molecular description, rather than the molar, in the future.

There is little difficulty in changing the units of rates. Thus y mol $\mathrm{l^{-1}}$ is clearly equivalent to $10^{-3}y$ mol $\mathrm{cm^{-3}}$ and also to $6.02 \times 10^{20}y$ molecules $\mathrm{cm^{-3}}$, giving the conversion factors for the rates.

In considering the rates of gas reactions, results are sometimes reported as mmHg per second at a reference temperature (not necessarily at the temperature of the experiments). These can be converted to mol $\mathrm{l^{-1}\,s^{-1}}$ by assuming that any gases present obey the ideal gas law, $PV = nRT$. Thus $n/V = P/RT$, and if n/V is to be in mol $\mathrm{l^{-1}}$ and P in mmHg, R must be in l mmHg $\mathrm{mol^{-1}\,K^{-1}}$ Thus z mmHg $\mathrm{s^{-1}}$ is equivalent to $z/62.32T$ mol $\mathrm{l^{-1}\,s^{-1}}$. Wherever possible, results for gas reactions should be given in concentration

---

* M. L. McGlashan, *Physicochemical quantities and units*, second edition p. 68. London : Royal Institute of Chemistry, 1971.

rather than pressure units because of possible confusion about the reference temperature.

Typical rate laws

The brief discussion in 1 of the collision theory and the transition-state theory of elementary reactions suggested that the rate of a simple reaction should be given, at constant temperature, by the equation

$$\text{rate} = k \times \prod_{\text{reactants}} (\text{concn})$$

where k is the *rate constant* of the reaction. Experiment has confirmed that a number of reactions between molecules or ions, the concentration of which can easily be determined during reaction, follow the expected rate equation or *rate law*. Thus the reaction

$$OH^- + CH_3Br \rightarrow CH_3OH + Br^- \quad .. \quad .. \quad (R4)$$

has a rate law $-d[OH^-]/dt = k_1[OH^-][CH_3Br]$ and in the early stages of the reaction $H_2 + I_2 \rightarrow 2HI$ (R2), before much product is formed, $-d[H_2]/dt = k_2[H_2][I_2]$, provided that in each case the temperature is held constant.

Inspection of these equations shows that if the concentration of each reactant was set equal to unity (*i.e.* one mole per litre) the rate of reaction would be numerically equal to the value of k. For this reason, the coefficient k has been called the *specific reaction rate*, but it is now more usually called the rate constant.

In these rate equations the rate is proportional to a product of two concentration terms, and they are called *second-order rate laws*. In each second-order law the rate constant, k, has particular dimensions. If the concentrations are in the usual units of mol l^{-1}, and time is in seconds, the equations show that

$$(\text{mol l}^{-1}\ \text{s}^{-1}) = k\ (\text{mol l}^{-1})\ (\text{mol l}^{-1})$$

and hence the dimensions of these rate constants are l mol^{-1} s^{-1}.

For each reaction a plausible transition-complex can be devised by bringing together two molecules, so the reaction is classed as *bimolecular*. For reaction R4 the transition complex is thought to be formed by approach of the OH^- ion along the C–Br axis, the methyl group passing through a planar structure in the complex and suffering inversion from halide to alcohol. There is evidence of this Walden inversion from experiments on optically active alkyl halides. A convenient nomenclature* that describes this

---

* *See* Gould, p. 251, 473 (further reading, p. 79), for description of the classification of substitution and elimination reactions.

molecular process is S_N2, indicating a substitution reaction of a nucleophilic reagent (OH^-) in a bimolecular step

$$OH^- + CH_3Br \rightarrow [HO\cdots\overset{H\;\;H}{\overset{\diagdown\diagup}{C}}\underset{H}{\overset{}{}}\cdots Br] \rightarrow HOCH_3 + Br^- \quad .. \quad (R4)$$

Reaction R2 has previously been treated by most investigators as a single-stage molecular reaction, but recent work has suggested (*see* footnote, p. 18) that it takes place in two stages, an equilibrated dissociation $I_2 \rightleftharpoons 2I$ being followed by reaction R2′

$$H_2 + 2I \rightarrow 2HI \quad .. \quad .. \quad .. \quad (R2')$$

Reaction R2′ is an elementary termolecular reaction, as three distinct entities meet to form the transition complex; this may have a geometry different from that postulated for R2—for example, it may be linear rather than square.

Reversible reactions

Many chemical reactions are reversible and proceed to an equilibrium state where the rate of the forward reaction equals that of the reverse reaction. It is necessary to take this into account when setting up the rate equations and determining the rates experimentally. Thus the experimental rate laws for R2 and R3, each determined under conditions where the other can be neglected, are

$$\text{(R2)} \quad H_2 + I_2 \rightarrow 2HI \qquad -d[H_2]/dt = k_2[H_2][I_2] \quad .. \quad .. \quad (6)$$

$$\text{(R3)} \quad 2HI \rightarrow H_2 + I_2 \qquad d[H_2]/dt = k_3[HI]^2 \quad .. \quad .. \quad (7)$$

In view of the discussion on page 18 it would be less satisfactory, though internally consistent, to write

$$\text{(R2)} \quad d[HI]/dt = k_2''[H_2][I_2] \quad (d[HI]/dt = -2d[H_2]/dt, \text{ so } k_2'' = 2k_2)$$

$$\text{(R3)} \quad -d[HI]/dt = k_3''[HI]^2 \quad (\text{and, for similar reasons, } k_3'' = 2k_3)$$

It would be entirely unsatisfactory to use k defined for R2 by $-d[H_2]/dt$ together with k'' defined for R3 by $-d[HI]/dt$.

In a mixture of all three gases, not at equilibrium, the net rate of formation of hydrogen will be given by the (algebraic) sum of eqns 6 and 7. Thus, in general

$$d[H_2]/dt = k_3[HI]^2 - k_2[H_2][I_2] \quad .. \quad .. \quad (8)$$

Notice that when starting with a mixture of H_2 and I_2, [HI] must be set equal to zero in eqn 8, whereupon eqn 6 is recovered; similarly, when starting from HI, $[H_2] = [I_2] = 0$ initially, and eqn 7 is recovered from eqn 8.

The two extreme conditions are the only simple ones for examining the reaction. Starting with hydrogen iodide, the gas can be run into a heated quartz vessel, held at a constant temperature near 300 °C, to give a chosen initial concentration. The mixture is withdrawn after a suitable lapse of time and rapidly chilled to prevent further reaction, thus allowing chemical analysis of the mixture present at the time of withdrawal. By repeating this at different intervals, it is found that the concentrations change smoothly with time. The rate can be derived by drawing tangents to the curve at any point; experiments show that, at low conversions, the rate is indeed proportional to $[HI]^2$. Next k_3 can be calculated by dividing the rate at any time by the appropriate value of $[HI]^2$. The rate law holds and k_3 is constant over a wide range of concentrations.

The reaction between H_2 and I_2 is much less easy to study because it is difficult to introduce high pressures of the iodine vapour. It has been accomplished, however, in sealed quartz vessels which were heated for a known period and then chilled and broken to extract the mixture for analysis. It was found that the rate followed eqn 6 in the initial stages and k_2 can be determined in the same way as k_3. It is harder to test eqn 8, which must be used later in either run, but it has been shown to fit the results when the values of k_2 and k_3, determined from the initial-rate experiments, are used.

At each temperature the reaction reaches equilibrium when the concentrations assume special values $[H_2]_e$, $[I_2]_e$ and $[HI]_e$ such that

$$k_2[H_2]_e[I_2]_e = k_3[HI]_e^2$$

Hence, transposing,

$$\frac{[HI]_e^2}{[H_2]_e[I_2]_e} = \frac{k_2}{k_3} = K_c \quad .. \quad .. \quad .. \quad (9)$$

where K_c is the equilibrium constant in terms of concentrations for the reaction. The experimental values of k_2/k_3 and of K_c are found to agree closely at different temperatures.

Rate constants and temperature

Experiments have shown that rate constants of most reactions are larger at higher temperatures. When this was first discovered attempts were made to fit the rate constants to equations containing simple powers of T. Arrhenius then showed that there were theoretical reasons for preferring exponential equations of the form

$$k = A \exp(-E/RT) \quad .. \quad .. \quad .. \quad (5)$$

He also showed that the results for some reactions fitted this equation with A and E characteristic of each reaction but independent of temperature. It was pointed out in chapter 1 that later theoretical treatments of elementary reactions predicted that A should vary slowly with temperature, but it is doubtful whether the predicted variations could be tested with the experimental data at present available.

Thus eqn 5 is usually adopted as a working hypothesis and is called the *Arrhenius equation.*

The energy E is the *activation energy*; when A is assumed independent of temperature, E controls the variation of the rate constant with temperature. The factor A is sometimes called the frequency factor of the Arrhenius equation, which suggests dimensions of s^{-1}. Inspection of eqn 5 will show that A has the same dimensions as k, and hence for the reactions being considered at present A has dimensions of $1\ mol^{-1}\ s^{-1}$. As will be seen later, k can have other dimensions in other reactions and so A is best called the *pre-exponential* or *Arrhenius factor.*

The Arrhenius equation predicts that k varies with temperature as in *Fig.* 4*a*. At very high temperatures, k will asymptotically approach A. At lower temperatures k falls, the rate of change being steepest at temperatures around $T = E/2R$. For most chemical reactions, k lies on the lower part of the curve, but for a few with very low activation energies the upper part of the curve can be realized at accessible temperatures.

Some typical evidence for the utility of eqn 5 for the simple reactions R2 and R3 is shown in *Fig.* 5. In this figure the results are plotted as $\log_{10} k$ against $1/T$. From eqn 5, by taking logarithms,

$$\ln (k/\text{units}) = \ln (A/\text{units}) - E/RT$$

and hence $$\log_{10} (k/\text{units}) = \log_{10} (A/\text{units}) - \frac{E}{2.303\ RT} \quad \cdot\cdot \quad (5')$$

The points give good fits to straight lines, showing that E and A are either independent of temperature or change in ways which compensate each other in eqn 5′. The first view is simpler and convenient, but should not be adopted dogmatically. There is evidence for compensation in some reactions, but at present only in complex and not in elementary ones.

It will be useful to point out here that a few reactions (*see* p. 31), which may be elementary, nevertheless show a rate constant decreasing as the temperature increases as in Fig. 4*b*. If the Arrhenius equation is fitted to these results, E appears to be negative, usually lying between 0 and -12 kJ mol^{-1}. The introduction of 'negative activation energies' has the practical advantage of

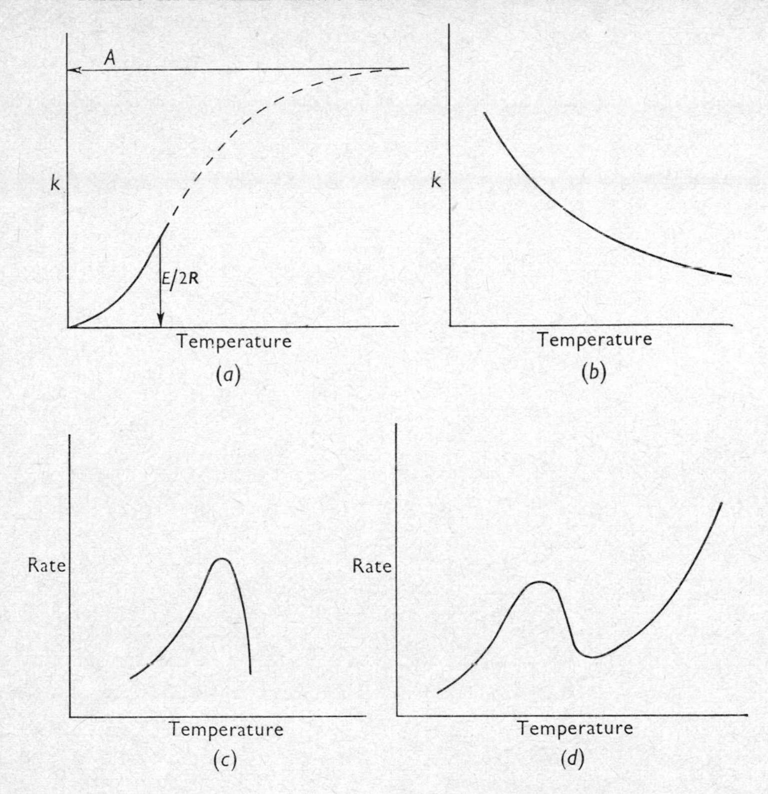

FIG. 4. Effect of temperature on rates: (*a*) Typical behaviour of rate constant of elementary reaction; (*b*) Rate constant of some third-order reactions; (*c*) Rate/temperature relationship with enzyme-catalysed reaction; (*d*) Rate/temperature relationship in oxidation of some hydrocarbons

describing the temperature-dependence of all elementary reactions by one simple equation, with reasonable accuracy relative to present experimental results. But it is hard to justify theoretically, and more likely causes of the behaviour of association reactions are discussed later (p. 32).

The rate constants of complex reactions may vary in a simple manner (*see Figs.* 4*a* or 4*b*) or in a complicated manner (*Figs.* 4*c* or 4*d*) as the temperature is raised (*see* p. 61).

It is not easy, without a lot of experience, to comprehend the effect of given values of E on k and hence on the rate. This is partly because a rate of reaction is not easily visualized quantitatively. An easier concept for showing the effect of E is the *half-life* ($\tau_{1/2}$) of a reaction, that is, the time taken for a reactant concentration to drop to one half of its original value. A fast

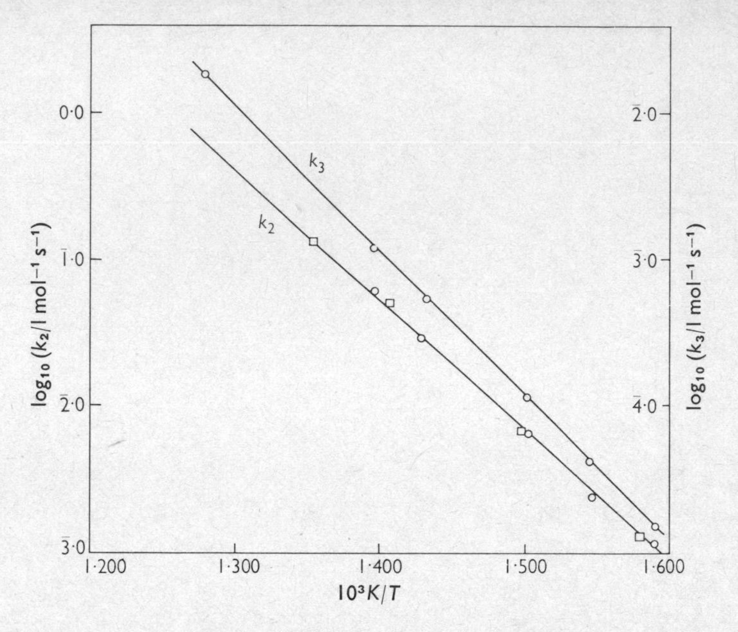

FIG. 5. Arrhenius plots for $2HI \xrightarrow{3} H_2 + I_2$ (upper line) and $H_2 + I_2 \xrightarrow{2} 2HI$ (lower line); k in $\mathrm{l\ mol^{-1}\ s^{-1}}$.
○ after Bodenstein (1899) □ after Sullivan (1959)

reaction obviously has a short half-life. The actual relationship between $\tau_{1/2}$, k and the original concentration varies with the type of the reaction (*see* p. 37) but for elementary reactions, under most experimental conditions, $\tau_{1/2}$ is inversely proportional to k.

In *Fig*. 6, values of $\tau_{1/2}$ are plotted against T on semi-logarithmic scales to contain the wide range of values of k for three reactions

(*a*) $Cl + ClNO \rightarrow Cl_2 + NO$ $A_a = 10^{10.1}\ \mathrm{l\ mol^{-1}\ s^{-1}}$, $E_a =$ 4 kJ mol^{-1}

(*b*) $H + D_2 \rightarrow HD + D$ $A_b = 10^{10.7}\ \mathrm{l\ mol^{-1}\ s^{-1}}$, $E_b =$ 31 kJ mol^{-1}

(*c*) $HI + HI \rightarrow H_2 + I_2$ $A_c = 10^{10.9}\ \mathrm{l\ mol^{-1}\ s^{-1}}$, $E_c =$ 185 kJ mol^{-1}

For the (atom + molecule) reactions, it is convenient to think of the half-life of the atoms when the molecular species is in great excess, say at a concentration equal to that in the gas at s.t.p. For the decomposition of hydrogen iodide the half-life can be calculated for an initial concentration equal to that at s.t.p.

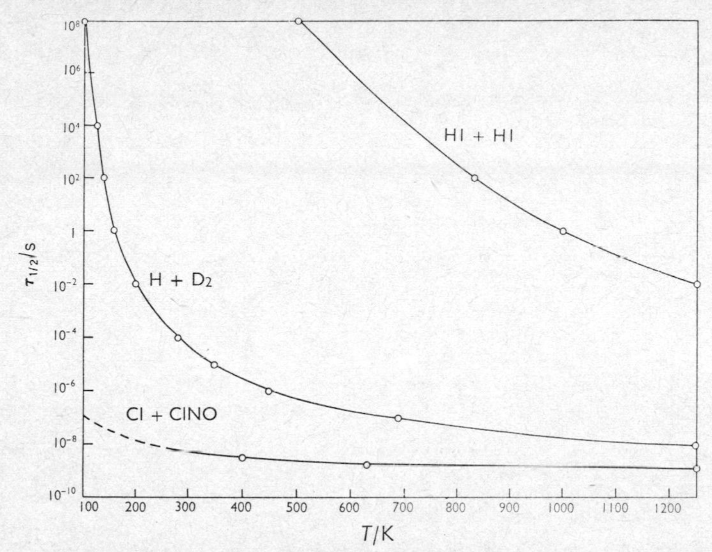

FIG. 6. Half-lives, $\tau_{1/2}$, for the reactions $2HI \rightarrow H_2 + I_2$, $H + D_2 \rightarrow HD + D$, $Cl + ClNO \rightarrow Cl_2 + NO$, at different temperatures

The following points should be noticed:

(*a*) For the reaction $Cl + ClNO \rightarrow NO + Cl_2$, with its very low activation energy, the half-life of the chlorine atoms is extremely short, about 10^{-9} s at 1000 K and about 5×10^{-9} s at room temperature. The NOCl liquefies at lower temperatures under 101 kPa pressure.

(*b*) For the reaction $H + D_2 \rightarrow HD + D$ the half-life is very short at room temperature, about 10^{-4} s, but it increases to 10^{-2} s at 'dry ice' temperature (about 200 K) and to about 10^8 s (3 years) at liquid-air temperatures (about 100 K).

(*c*) For the reaction $2HI \rightarrow H_2 + I_2$ the half-life at room temperature is very long, about 10^{17} s (or 3×10^9 years!) but at 500 K it has dropped to 10^8 s (3 years), at 800 K to 100 s and at 1000 K to 1 s.

Activation energies of reversible elementary reactions

Equations like (9) (p. 24) hold for reversible elementary reactions. In general the general relationship between the rate constant k_f of the forward reaction, the rate constant k_b of the reverse reaction, and the equilibrium constant K_c of reversible elementary reactions is:

$$K_c = k_f/k_b$$

Hence, for such reactions, $\ln(K_c/\text{units}) = \ln(k_f/\text{units}) - \ln(k_b/\text{units})$

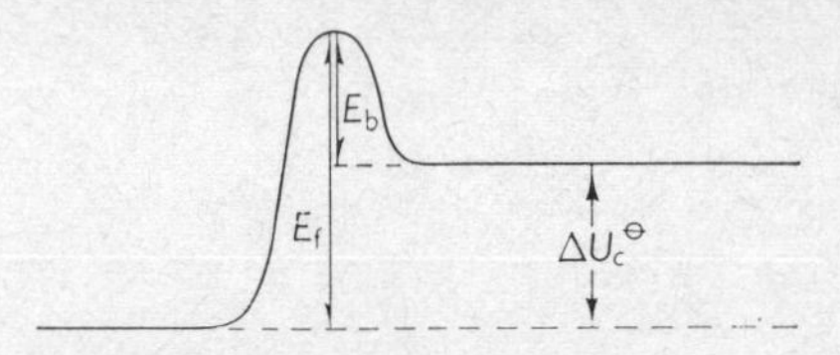

FIG. 7. Relationship between internal energy change, $\Delta U^{\ominus}_{c}$, and the activation energies, E_{f} and E_{b}, for a reversible elementary reaction

and
$$\frac{\mathrm{d}\ln K_c}{\mathrm{d}T} = \frac{\mathrm{d}\ln k_{\mathrm{f}}}{\mathrm{d}T} - \frac{\mathrm{d}\ln k_{\mathrm{b}}}{\mathrm{d}T}$$

Now $\mathrm{d}\ln K_c/\mathrm{d}T = \Delta U^{\ominus}_{c}/RT^2$, from the isochore, and from eqn 5 $\mathrm{d}\ln k_{\mathrm{f}}/\mathrm{d}T = E_{\mathrm{f}}/RT^2$ and $\mathrm{d}\ln k_{\mathrm{b}}/\mathrm{d}T = E_{\mathrm{b}}/RT^2$. Therefore

$$\Delta U^{\ominus}_{c} = E_{\mathrm{f}} - E_{\mathrm{b}}$$

This relationship holds for the (hydrogen + iodine) reaction; indeed, wherever it has been possible to test it, it holds, within the accuracy with which the activation energies can be measured. It is entirely in keeping with the idea of a common transition-complex for reversible elementary reactions, as can be seen from the energy profile for a reaction when the energies are for one mole of transition complex rather than one molecule (*Fig.* 7). Another important point is that the activation energy of an endothermic reaction must be equal to, or greater than, the endothermicity of the reaction. A high absorption of heat in a reaction, as in many dissociations or decompositions, means a high activation energy, and this in turn means that the reaction is bound to be slow at low temperatures. For such a reaction, however, the rate will increase rapidly if the temperature is raised.

Second-order, bimolecular reactions

Some typical reactions that are thought to be elementary second-order, bimolecular reactions are listed in Table 1.

With the gas-phase reactions of small molecules or atoms the A factors are all high and close to the maximum values predicted by the collision theory. The reactions of larger molecules have much smaller A factors. This cannot be satisfactorily explained by the collision theory, as it seems very unlikely that steric demands would cause P factors as low as 10^{-4} or 10^{-5}. The transition theory successfully explains the lower values by the greater loss of 'storage capacity' for rotational energy which decreases the entropy change, $\Delta S^{\ominus}_{c}$, in forming the transitional complex. Now,

$$\frac{\ln K_c}{\text{units}} = -\frac{\Delta G_c^\ominus}{RT} = -\frac{\Delta H_c^\ominus}{RT} + \frac{\Delta S}{R}$$

and hence a smaller $\Delta S_c^\ominus$ means a smaller K_c and a slower rate of reaction (eqn 2, p. 13).

A similar pattern can be observed for reactions in solution. For the smaller reactants the A factors are close to those shown for the gas phase, and they are much lower for the association reactions between large molecules.

The activation energies are higher when several bonds have to be stretched in forming the transition complex. The effect of different activation energies on the value of k and the half-life of the reaction has already been discussed.

Table 1

	$\frac{A}{\text{l mol}^{-1}\,\text{s}^{-1}}$	$\frac{E}{\text{kJ mol}^{-1}}$
Gas-phase molecular reactions*		
$2HI \rightarrow H_2 + I_2$	9.0×10^{10}	185
$2NOCl \rightarrow 2NO + Cl_2$	5.0×10^{9}	99
$2NO_2 \rightarrow 2NO + O_2$	3.9×10^{9}	116
Gas-phase atom + molecule reactions		
$H + D_2 \rightarrow HD + D$	5.2×10^{10}	31
$Cl + NOCl \rightarrow NO + Cl_2$	1.1×10^{10}	4
$Cl + H_2 \rightarrow HCl + H$	8.0×10^{10}	23
Gas-phase dimerizations		
$2\,CH_2{=}CH{\cdot}CH{=}CH_2 \rightarrow$ H_2 C CH CH·CH=CH_2 ‖ | CH CH_2 C H_2	4.7×10^{7}	106
$CH_2{=}CH{\cdot}CH{=}CH_2 + CH_2{=}CH{\cdot}CHO \rightarrow$ H_2 C CH CH·CHO ‖ | CH CH_2 C H_2	1.5×10^{6}	83
Aqueous solution substitution reactions		
$CH_2I{\cdot}COO^- + OH^- \rightarrow CH_2OH{\cdot}COO^- + I^-$	6.3×10^{11}	94
$CH_3Br + I^- \rightarrow CH_3I + Br^-$	1.7×10^{10}	76
$CH_3I + OH^- \rightarrow CH_3OH + I^-$	1.2×10^{12}	93
Additive reactions in solution		
$(C_2H_5)_3N + C_2H_5Br$ (in C_6H_6)	3.1×10^{2}	47

* There are alternative two-stage paths for the decompositions of NOCl and NO_2 proceeding in parallel with the molecular reactions shown (*see* p. 63). The initial products of reaction of two HI molecules may be H_2 + 2I, with $2I \rightleftharpoons I_2$ (*see* p. 18).

First-order, unimolecular reactions

There is no reason *a priori* why a reaction which involves bond-breaking and bond-making in only one reactant molecule should not proceed in a single step, provided a suitable transition state can be visualized. This would be a *unimolecular reaction*. The rate of such a step would be proportional to the concentration of the reactant, so giving a *first-order* rate law. The decomposition of cyclobutane, the isomerization of cyclopropane, and the isomerization of vinyl allyl ether in the gas phase follow the expected rate law at pressures above 101 kPa.

Table 2

	A/s^{-1}	E/kJ mol^{-1}
H_2C——CH_2 \ / C H_2 $\rightarrow CH_3{\cdot}CH{=}CH_2$	1.6×10^{15}	272
H_2C—CH_2 | | H_2C—CH_2 $\rightarrow 2C_2H_4$	4.0×10^{15}	262
$CH_2{=}C(H)$—O—CH_2—$CH{=}CH_2$ $\rightarrow$ CH_2—$C(H){=}O$, CH_2—$CH{=}CH_2$	1.8×10^{11}	128

For each of these reactions the rate law is of the form

$$-\frac{dc}{dt} = kc$$

where c is the concentration of the reacting species. Consideration of the dimensions of each side of this equation shows that k must be expressed as time^{-1} (usually s^{-1}) for each of these reactions. The rate constants also obey the Arrhenius equation and values of A and E are given in Table 2.

The activation energies for the decompositions are usually high, as more bonds are broken than are formed. For some rearrangements the activation energies are not so high, usually because a comparable number of bonds are broken and formed.

Many other examples are now known, but it is also known that many first-order reactions at one time thought to be elementary are in fact complex. Thus the decomposition

$$2N_2O_5 \rightarrow 2N_2O_4 + O_2$$

follows a first-order rate law and was for long thought to be unimolecular. However, it is now known to be complex (*see* p. 56); this might be expected from the impossibility of making a transition complex that can produce the products directly from one molecule.

Where elementary unimolecular reactions have been studied in the gas phase and in solution they have very nearly the same rates in both phases.

However, with these and similar reactions in the gas phase the rate law changes from first to second order as the pressure is decreased. The most probable explanation of this behaviour is that at lower pressures the rate is determined by the rate of production, in binary collisions, of the activated transition-state. This has great significance for the theories of elementary reactions (*see* p. 54f.).

Third-order, termolecular reactions

There are only a few reactions in which three molecules take part and a third-order rate law is observed in keeping with the chemical equation. It is possible to visualize a transition complex for the reactions of nitric oxide with oxygen, chlorine or other halogens, and it is probable that these reactions are true elementary termolecular reactions. However, it is also possible that they proceed in two stages and are really complex reactions. The same is true of the reaction of ethylene oxide with aqueous hydrogen chloride or bromide. The condensation of benzaldehyde to benzoin, catalysed by cyanide ion, is also third order but is probably complex. The reaction between hydrogen and iodine may involve a termolecular step (*see* p. 18).

$$2NO + O_2 \rightarrow ON\cdots O\cdots O\cdots NO \rightarrow ONO + ONO$$

$$2NO + Cl_2 \rightarrow ON\cdots Cl\cdots Cl\cdots NO \rightarrow ONCl + ClNO$$

$$\overset{\displaystyle O}{CH_2\text{—}CH_2} + H^+ + Br^- \rightarrow \overset{\displaystyle H \atop O \quad Br}{CH_2\text{—}CH_2} \rightarrow CH_2(OH)CH_2Br$$

These reactions have rate laws of the dimensional form

$$-\frac{\mathrm{d\ concn}}{\mathrm{d}t} = k(\mathrm{concn})^3$$

and so their rate constants are expressed in units of concentration^{-2} time^{-1}, usually as l^2 mol^{-2} s^{-1}. For the second reaction the rate constants follow an Arrhenius Law with $E = 25$ kJ mol^{-1}, $A = 10^3$ l^2 mol^{-2} s^{-1}. The first reaction is exceptional, however. Its velocity constant *decreases* as the temperature is raised from -190 °C to 500 °C; at the higher temperature it is practically constant (*see Fig.* 4*b*, p. 25). One explanation of this suggests that the reaction is complex, proceeding in the two stages:

$$2NO \rightleftharpoons (NO)_2$$

$$(NO)_2 + O_2 \rightarrow 2NO_2$$

where $(NO)_2$ is a true intermediate and not the transition complex. The transition-state theory, however, explains the effect by assuming a single termolecular reaction with very low activation energy; with the transition state formed in this way, the Arrhenius factor A is proportional to $T^{-7/2}$ and this inverse temperature-dependence can account for the fall of k when the temperature is raised.

As already explained (*see* p. 6) the recombination of free atoms in the gas phase is a third-order reaction, as some third body is required to take away the energy released by the formation of the molecule, *e.g.*

$$Br + Br + M \rightarrow Br_2 + M^*$$

For these reactions, the velocity constant also decreases as the temperature is raised. This behaviour can also be explained by the transition-state theory, although the Arrhenius factor A now depends on a smaller negative power of the temperature than with the reaction $2NO + O_2 \rightarrow 2NO_2$. It is also possible that the reaction proceeds through two stages, such as

$$Br + Br \rightleftharpoons Br_2^*$$

$$Br_2^* + M \rightarrow Br_2 + M^*$$

or, especially with certain molecules as M, through the alternative sequences

$$Br + M \rightleftharpoons MBr$$

$$MBr + Br \rightarrow Br_2 + M$$

Species forming the transition complex

In these elementary reactions the requisite number of reactant molecules come together to form the transition complex. The order of the rate law and the molecularity of the reaction are identical. The experimental rate law specifies the number of molecules of reactants which are needed to form the transition complex, and so indicates the composition of the complex.

Ambiguities can arise, however, if equilibria exist between the reactants and other species. It is possible that the transition complex is formed from some intermediate species rather than the original reactants. Such a situation is uncommon in the gas phase, but fairly common in solution—for example, as a result of hydrolysis of ions in aqueous solution or complex formation. An example that has been studied extensively is the formation of urea from ammonium cyanate. The rate law is

$$-\frac{d[NH_4CNO]}{dt} = k[NH_4^+][CNO^-]$$

However, because of the rapid equilibria

$$NH_4^+ + CNO^- \rightleftharpoons NH_3 + HCNO$$

and

$$HCNO \rightleftharpoons HNCO$$

it is possible* that NH_3 and HNCO are the immediate precursors of the transition complex. They give a more likely transition complex than could easily be formed from the ions, thus:

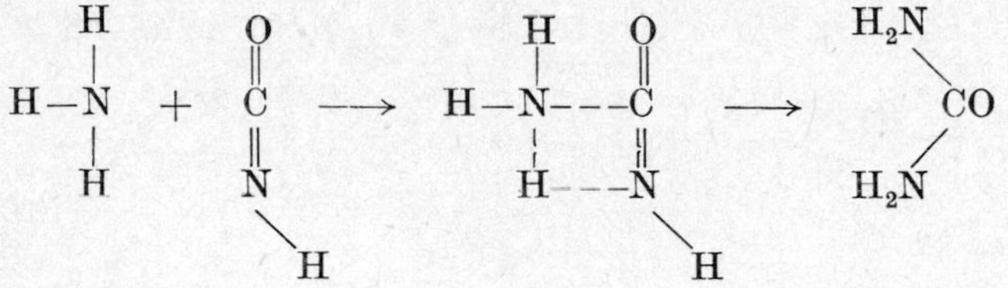

```
   H          O              H   O                 H2N
   |          ||             |   ||                   \
H—N    +     C    ——>    H—N---C      ——>            CO
   |          ||             ¦   ||                   /
   H          N              H---N                 H2N
               \                  \
                H                  H
```

In cases like this, it proves impossible to say from kinetic measurements which pair of potential reactants actually form the complex. If, however, a previous equilibrium exists of the form

$$A + B \rightleftharpoons C + D$$

and only *one* of the species present forms the transition complex (*e.g.* by reaction with another species E), it proves possible to detect the pre-equilibria and the species forming the complex. Several examples of this will be given while considering complex rate laws.

Integration of simple rate laws

There are four main *purposes* in deriving the integrated forms of simple rate laws:

(1) to allow the calculation of concentration, c_t, of a reactant at time t, given the concentration c_0 at $t = 0$, the particular rate law and the appropriate value of k;

(2) to provide additional tests to show which rate law fits certain experimental data;

---

* *See* Frost and Pearson, Chapter 11B (further reading, p. 79).

(3) to derive a rate constant, k, from experimental data, having established the rate law for the reaction;

(4) to obtain expressions relating the fractional lives (*e.g.* half-lives) to k and c_0.

We shall not attempt to examine the integrations in any detail, as the methods are clearly given in the textbooks listed (p. 79). But it is important to emphasize the *general* procedure, which should be followed for all reactions, no matter how simple the rate aw or the chemical equation:

(*a*) write down the (overall) chemical equation with the concentrations at time 0 and time t set underneath each reactant and product;
(*b*) write down the rate equation and substitute for each concentration;
(*c*) carry out the appropriate integration, establishing the integration constant by using the initial concentrations at $t = 0$;
(*d*) modify the integrated equation for testing, if necessary.

The decomposition of cyclopropane (CP) will be considered as a very simple example of this procedure

(*a*)

$$\underset{\underset{H_2}{C}}{H_2C—CH_2} \xrightarrow{k_1} CH_3{\cdot}CH{=}CH_2$$

Initial concentration	$[CP]_0$	0
Concentration at time, t	$[CP]$	$[CP]_0 - [CP]$

(*b*)

$$-\frac{d[CP]}{dt} = k_1[CP]$$

(*c*)

$$-\int \frac{1}{[CP]}\, d[CP] = \int k_1 dt$$

$$-\ln [CP] = k_1 t + \text{constant}$$

$$\text{At time 0, } -\ln [CP]_0 = \text{constant}$$

$$\therefore \ln [CP]_0 - \ln [CP] = k_1 t$$

Equivalent forms of this equation are

$$\ln \frac{[\mathrm{CP}]_0}{[\mathrm{CP}]} = k_1 t$$

$$\text{or } [\mathrm{CP}] = [\mathrm{CP}]_0 . \mathrm{e}^{-k_1 t}$$

(*d*) The most convenient form of the expression is

$$\ln \frac{[\mathrm{CP}]}{\text{units}} = \ln \frac{[\mathrm{CP}]_0}{\text{units}} - k_1 t$$

Thus a plot of ln [CP]/units against t should be linear, with a negative slope numerically equal to k_1. The units of k_1 will be (unit chosen for time $t)^{-1}$.

The standard forms of the common rate laws for elementary reactions, the integrated equations, the simplest functions for graphical plotting and the expressions for the half-lives (see next section) are given in Table 3.

One more example will be given to illustrate the treatment of a reaction which will be discussed again on p. 47. This is the oxidation of iodide ion by the persulphate ion; it is a complex reaction but has a simple rate law.

(*a*)	$S_2O_8^{2-}$ +	$2I^-$	$\rightarrow 2SO_4^{2-}$ +	I_2
Initial concentration	$[\mathrm{A}]_0$	$[\mathrm{B}]_0$	0	0
Concentration at time, t	$[\mathrm{A}]_0 - x$	$[\mathrm{B}]_0 - 2x$	$2x$	x

(*b*) The experimental rate law is $\dfrac{\mathrm{d}[I_2]}{\mathrm{d}t} = k[S_2O_8^{2-}][I^-]$

$$\text{Hence } \frac{\mathrm{d}x}{\mathrm{d}t} = k([\mathrm{A}]_0 - x)([\mathrm{B}]_0 - 2x)$$

(*c*) This integrates to give: $\ln \dfrac{[\mathrm{B}]}{[\mathrm{A}]} = ([\mathrm{B}]_0 - 2[\mathrm{A}]_0)kt + \ln \dfrac{[\mathrm{B}]_0}{[\mathrm{A}]_0}$

Notice that this is very similar to the standard form (v) in Table 3 with a different multiple for kt.

(*d*) This can be tested directly by plotting $\ln \dfrac{[\mathrm{B}]}{[\mathrm{A}]}$ against t.

Fractional lives

The half-life, $\tau_{1/2}$, or any other fractional life is unambiguous for elementary first-order reactions when the reverse reaction is

Table 3. Some standard integrated equations and expressions for half-lives

Type of reaction	Initial conditions	Rate equation	Integrated equation	Simplest plot	Half-life $\tau_{1/2}$
(i) A → products	$[A] = [A]_0$	$-\frac{d[A]}{dt} = k_0$	$[A] = [A]_0 - k_0 t$	$[A]/t$	$\frac{[A]_0}{2k_0}$
(ii) A → products	$[A] = [A]_0$	$-\frac{d[A]}{dt} = k_1[A]$	$\ln\frac{[A]}{\text{units}} = \ln\frac{[A]_0}{\text{units}} - k_1 t$	$\ln\frac{[A]}{\text{units}}/t$	$\frac{\ln 2}{k_1} = \frac{0.69}{k_1}$
(iii) 2A → products	$[A] = [A]_0$	$-\frac{d[A]}{dt} = k_2[A]^2$	$\frac{1}{[A]} = \frac{1}{[A]_0} + k_2 t$	$\frac{1}{[A]}\Big/t$	$\frac{1}{k_2[A]_0}$
(iv) A + B → products	$[A]_0 = [B]_0$				
(v) A + B → products	$[A]_0 \neq [B]_0$	$-\frac{d[A]}{dt} = k_2'[A][B]$	$\ln\frac{[B]}{[A]} = \ln\frac{[B]_0}{[A]_0} + ([B]_0 - [A]_0)k_2' t$	$\ln\frac{[B]}{[A]}\Big/t$	Not useful
(vi) A + B + C → products	$[A]=[B]=[C]$	$-\frac{d[A]}{dt} = k_3[A][B][C]$	$\frac{1}{[A]^2} = \frac{1}{[A]_0^2} + k_3 t$	$\frac{1}{[A]}\Big/t$	$\frac{3}{k_3[A]_0^2}$
(vii) 2A + B → products	$[A]_0 = 2[B]_0$	$-\frac{d[B]}{dt} = k_3'[A]^2[B]$	$\frac{1}{[B]^2} = \frac{1}{[B]_0^2} + 4k_3' t$	$\frac{1}{[B]}\Big/t$	$\frac{3}{4k_3'[B]_0^2}$

Notes: (1) The expressions for third-order reactions of the type A + B + C → products, with $[A]_0 \neq [B]_0 \neq [C]_0$, and for the type 2A + B → products, with $[A]_0 \neq 2[B]_0$, should be treated as special cases.

(2) There is little point in extending the table to reactions of higher order.

(3) Zero-order reactions are always complex processes: k_0 usually includes other concentration terms.

negligible. Thus for the reaction

$$\begin{array}{c} H_2C\text{—}CH_2 \\ \diagdown\ \diagup \\ C \\ H_2 \end{array} \xrightarrow{k_1} CH_3{\cdot}CH{=}CH_2$$

the rate law is

$$-\frac{d[CP]}{dt} = k_1[CP]$$

so that

$$\ln\frac{[CP]_0}{[CP]} = k_1 t$$

and

$$\ln 2 = k_1\tau_{1/2}$$

$\therefore \tau_{1/2} = \dfrac{\ln 2}{k_1}$ and $\tau_{1/2}$ is independent of the initial concentration.

It is also unambiguous for a second-order reaction in which one reactant is in great excess. Thus in the reaction

$$Cl + ClNO \xrightarrow{k_2} Cl_2 + NO$$

the reverse reaction can be ignored at low and moderate temperatures, and $[Cl] < [ClNO]$. The reaction is pseudo first-order in [Cl], and *for the atoms*

$$\tau_{1/2} = \frac{\ln 2}{k_2[ClNO]}$$

Thus the half-life depends on the nearly constant value of [ClNO] but not on the initial concentration of atoms.

For simple second- and third-order reactions, where the reactants are mixed in the proportions in which they react and the reverse reactions are negligible, the half-lives depend on the velocity constants and the initial concentrations, as shown in the last column of Table 3.

If the reactants are mixed in proportions other than those in which they react, or the reactions are reversible, the concept of a half-life becomes much less useful. A reactant in considerable excess may not have a finite half-life, and for other conditions the expression may be very cumbersome. If the reactions are reversible, the half-life may be calculated for (say) the forward reaction, ignoring the reverse; but it gives no information about the time taken to get (say) half-way to the equilibrium position. The time taken for a process to move from a 'disturbed' or non-equilibrium position towards the equilibrium position, by a definite fraction $1/e$ of the original distance from equilibrium, is the *relaxation time* of the process. It is especially useful for first-order processes, because it is independent of the absolute concentrations. It is often used in studies of very fast reactions,* or of the exchange of translational and vibrational energy, or of the time taken for chain centres to reach their steady-state concentration.

Experimental measurement of reaction rates

We have already outlined (pp. 22 and 23) the methods of measuring the rates of the gas-phase molecular reactions

$$H_2 + I_2 \rightarrow 2HI$$
$$2HI \rightarrow H_2 + I_2$$

and later we shall indicate how some reactions in solution can easily be followed:

$$S_2O_8^{2-} + 2I^- \longrightarrow 2SO_4^{2-} + I_2 \quad \text{(p. 47)}$$
$$H_2O_2 + 2I^- + 2H^+ \longrightarrow 2H_2O + I_2 \quad \text{(p. 48)}$$
$$BrO_3^- + 5Br^- + 6H^+ \longrightarrow 3Br_2 + 3H_2O \quad \text{(p. 48)}$$

Some more specialized techniques have previously been mentioned when discussing the reactions

$$Br_2 + M \longrightarrow 2Br + M \quad \text{(p. 4)}$$
$$Br + Br + M \longrightarrow Br_2 + M \quad \text{(p. 7)}$$
$$H + D_2 \longrightarrow HD + D \quad \text{(p. 9)}$$

These necessarily brief accounts indicate the wide range of techniques used to measure reaction rates. There are, however, simple principles that apply to nearly all measurements of rates. Thus most† experimental measurements of reaction rates involve preparing the reactants, mixing them at the desired temperature, and subsequently measuring changes of concentration of reactants and products as early and then for as great an extent of reaction as possible. The experiments usually give the values of the concentrations of one or more species at various times after mixing, and the rates are determined from the slopes of plots of concentration against time. Manual plotting and tangent-drawing are not very accurate, and some improved methods have been described (*see, e.g.*, Friess and Weissberger, pp. 180; further reading, p. 79).

* *See*, for example, E. F. Caldin, *Fast Reactions in Solution*. Oxford: Blackwell Scientific Publications, 1964.

† There are a few exceptions when an equilibrium mixture is set up and then disturbed, and the time is measured for the system to 'relax' to the new equilibrium position.

There are two basic schemes for mixing the reactants and following the reaction. In the *static* method, chosen quantities of the reactants are mixed for each experiment in the reaction vessel whose temperature must be kept as constant as possible. For reactions in solution the separate reactants are brought to the temperature of the experiment and then mixed; if they were mixed at a lower temperature, their high heat capacities would lead to a long interval of rising temperature. With gaseous reactants, of low heat capacity, it is often satisfactory to make a mixture at room temperature and admit it to a heated vessel; obviously this can only be done if the reaction is much slower at lower temperatures. After mixing, the changes in concentration are followed by direct analysis of one or more chemicals present. There are considerable advantages in analysing more than one chemical. Many important changes in our knowledge of the course of various reactions have come through the application of more discriminating methods of analysis; ideally an independent determination of each chemical is required. The use of chemical analysis, chromatography, mass spectrometry and absorption spectroscopy or photometry are most desirable. In some cases, however, particularly in the first studies of a rate, some overall property such as pressure, volume or thermal or electrical conductivity is often used. The analysis may be continuous if it is based on some physical property, or it may involve intermittent sampling. The latter has proved convenient with solutions, for a sample can be drawn off without affecting the concentrations of the remaining chemicals. It is less satisfactory for static gas-mixtures, because the removal of a sizeable sample alters the concentrations of the components of the mixture in a fixed volume. It can be used, however, if the methods of analysis are sufficiently sensitive, as for example with recent developments in gas–liquid chromatography, to allow the withdrawal and analysis of very small samples. The reaction in a liquid or gaseous sample is often 'quenched' by removing an essential reactant chemically, by rapid cooling or by rapid dilution.

The second scheme for mixing the reactants and following the reaction is to use a *flow system*. In this method the reactants, usually preheated to reaction temperature, mix and flow continuously through a reaction vessel (the flow is usually axial along a tubular vessel), thus spreading the reaction in space rather than time. Samples for analysis can be withdrawn at intervals down the tube, or *in situ* analyses made by physical methods. From the known rates of flow the spatial intervals are readily converted to time intervals. These systems have several important advantages over static systems. Mixing can be made very rapid, without the danger present in static systems of adiabatic heating of gaseous reactants to temperatures higher than that required. Relatively large amounts of products can be collected at early stages of the reaction, provided that the samples can be quenched successfully. Both these features are helpful when investigating rapid reactions. On the other hand, for gas-phase reactions with a change in the number of molecules as the reaction proceeds, the changes of partial pressure down the flow tube are complex; also, with high rates of flow there is some danger of 'plug' flow down the centre of the tube through a sheath of stagnant gas or solution. Flow systems have often been used in investigations of industrial processes, and there has recently been a renewed interest in them for following fast reactions in solution (*e.g.* acid–base reactions, enzyme-catalysed reactions and the formation of complexes) and in the gas phase when the reacting species are in low concentrations relative to some carrier gas (*e.g.* free-atom reactions of O, H and N at low pressures).

There is an interesting variation of the flow method, known as the '*stirred reactor*' *method*. With a selected rate of flow of reactants into the stirred reactor the rate of outflow of products is measured. The uniform composition of the mixture then produces products at a rate equal to the rate of outflow. Changes in the composition in the reactor change the rate of outflow of products, so that the dependence of rate on composition can be determined (*see* Friess and Weissberger, p. 66).

The general methods mentioned so far can be applied to both homogeneous and heterogeneous reactions. It is rare for a heterogeneous reaction to be followed except through changes in the bulk composition of a phase in contact with the surface, although some studies of chemisorption of gases on metals have utilized changes in properties of the metal such as electrical conductivity, surface accommodation coefficient or the work function for emission of electrons. With heterogeneous reactions, there are special problems of ensuring ready access of the reactants to the surface, particularly if this is the internal surface of a porous solid, and corresponding problems of representative sampling.

There are many special methods for following rates that cannot be dealt with here. For the use of radioactive tracers for following individual steps in complex reactions, and the use of competitive reactions for determining rates, *see* Friess and Weissberger, ch. 8.

It is extremely important that the reaction should be carried out at a uniform known temperature. Much attention has been paid to the design of thermostats and to the measurement of temperatures of vessels and reactants. There is particular difficulty with temperature gradients if very exothermic or endothermic gas reactions are being studied in large vessels. It is noteworthy that many of the reactions on which the classical studies of reaction kinetics were based were catalysed reactions in solution that could proceed at a reasonable rate at, or near, room temperature, when control of temperature is easy. The range of temperature that can be investigated depends upon factors such as the extremes of rates that can be measured and the onset of side-reactions or decomposition of reactants or solvent at higher temperatures. It is clearly more limited for reactions in solution, but fortunately the rates and temperatures can then be measured rather precisely. The range for gas reactions has steadily increased, and now extends to above 2000 °C with some studies of reactions in chemical shock-tubes at slightly lower temperatures; flames themselves have been used as hot vessels.

Establishment of rate laws from experimental data

The experimental information consists of sets of values of the concentrations of reactants and products at various times after mixing. If the reaction is thought to be elementary, the procedure is straightforward. The appropriate rate law (Table 3) is tested by plotting the results in the form suggested for the integrated equation. If the correct law has been chosen and the reaction is elementary, a linear plot should be obtained for at least 90 per cent of the reaction. Of course, the experimental errors in determining concentrations or time in any one experiment will lead to some scatter around a straight line, but this should show no steady trend away from the line if a number of experiments are carried out under uniform conditions (especially of temperature).

It is useless to draw conclusions from one run where the reaction is followed for a time much less than the half-life. The danger of this is illustrated by *Fig.* 8. A reaction has been examined to 10 per cent conversion, and the results are first plotted as 1/[A] versus time because the rate law is believed to correspond to reactions (iii) or (iv) in Table 3 (p. 36). This plot is shown in *Fig.* 8*a* (by ○ and left-hand scale). However, if $\log_{10}$ [A]* is plotted against time, as shown in *Fig.* 8*a* (by □ and right-hand scale), a very tolerable straight line is obtained! It is impossible even with accurate experimental results to select the rate law from this *one short experiment.* Another experiment was carried out, with double the initial concentration, but again restricting conversion to less than 10 per cent; the corresponding plots in *Fig.* 8*b* were obtained. Comparison with *Fig.* 8*a* reveals that the reciprocal plots have the same slope, but the slope of the log plots are in the ratio 2:1—in other words, the 'apparent first-order constants' are proportional to the initial concentrations and the reaction is really second-order, as confirmed by the two reciprocal plots of identical slope.

* strictly, $\log_{10}$ ([A]/units)

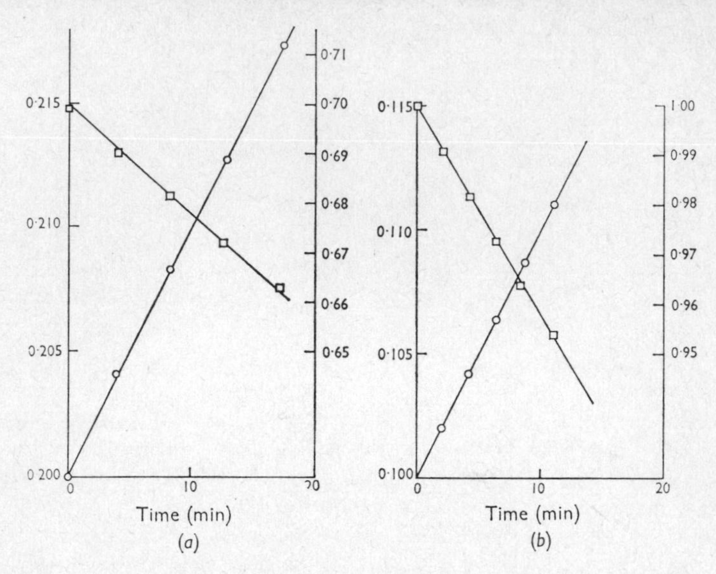

FIG. 8 (*a*) Plots of $1/[A]$ versus time (○ and left-hand scale) and $\log_{10}([A]/\text{units})$ versus time (□ and right-hand scale) for early stages of reaction $aA \rightarrow$ products; (*b*) Similar plots for double the initial concentrations of A

The same conclusion would be obtained if the initial experiment had been carried out to conversions well beyond 50 per cent.

The principles that must be adopted when designing experiments to test rate laws are:

(1) carry out enough experiments to check the reproducibility of results;

(2) carry the reaction as nearly to completion as possible when testing plots for linearity; *or*

(3) carry out experiments with widely different initial concentrations of reactants.

If the rate of reaction falls off much more rapidly than expected from any of the simple rate laws it may be elementary but reversible. The obvious tests here are to examine the effect of adding the products, or to investigate the initial stages of the reverse reaction, as suggested earlier (p. 22) for the reactions

$$H_2 + I_2 \rightleftharpoons 2HI$$

It is not easy to lay down standard procedures for examining a reaction which is suspected to proceed by a complex many-stage mechanism. Generally speaking, for a reaction

$$aA + bB + cC \rightarrow \text{products}$$

it is first assumed that the rate law is reasonably simple, and of the form

$$\text{Rate} = k[A]^{\alpha} [B]^{\beta} [C]^{\gamma} \ldots$$

where α, β, γ etc. are likely to be equal to or less than the corresponding coefficients a, b, c etc. By choosing special reactant conditions, α, β, γ *etc.* can be determined.

(*a*) *The initial-rate method.* The changes of concentration are measured at the beginning of the reaction in order to obtain the initial rate as accurately as possible at $t = 0$ when the concentrations c_A, c_B etc. are known. If the concentration of only one reactant is varied from run to run, by a factor x, then we have

$$(\text{Initial Rate})_1 = k.c_A^{\alpha}.c_B^{\beta}.c_C^{\gamma} \ldots$$

$$(\text{Initial Rate})_2 = k.(xc_A)^{\alpha}.c_B^{\beta}.c_C^{\gamma}$$

$$\therefore \quad \frac{(\text{Initial Rate})_2}{(\text{Initial Rate})_1} = x^{\alpha}$$

In this way α can be determined. By changing each reactant concentration in turn, the other indices β, γ, etc. can be determined. It is essential to examine as wide a range of values of x as possible. This is a powerful method, limited in accuracy by the experimental difficulties of determining initial rates.

(*b*) *The isolation method.* In this method the concentrations of all but one reactant are kept high, so as to be nearly constant during a run. The rate will then vary during the run because of changes in the concentration of the one 'isolated' reactant, and the dependence can be determined as described for a single reactant. Each reactant is isolated in turn, and the complete rate law established. This method is limited, however, by the need to keep the excess concentrations low enough to give rates that can be measured.

Of course, these two special methods can also be applied to elementary reactions and are often used in preference to selecting a likely rate law from Table 3.

More complex rate laws are unfortunately common, but must be examined with some likely mechanism in mind as indicated in Chapter 3.

3. Rate Laws for some Complex Reactions

On examining a wide range of reactions it becomes apparent that it is rare for a whole reaction to behave as simply as do those described so far. The rate law can rarely be predicted from the equation for a chemical reaction. Sometimes the rate law is more complex than would be expected from the chemical equation; often it is simpler. We shall examine some unexpected rate laws, and show that although puzzling at first sight, they can be rationalized by applying some simple principles. As these principles have been successful wherever they have been applied, it is well to state them immediately. First, it is assumed that most reactions do not proceed in one single stage, but in a series of stages or *reaction steps*. In the early ones, some or all of which may be reversible, intermediate species are produced. These may be molecules, ions, free radicals or atoms; later they react to yield the products. Secondly, each reaction step behaves as an elementary reaction with its own transition complex, and its rate can be predicted from the chemical equation for the step by setting it proportional to the concentration of each reactant in the step. Thirdly, the overall rate is controlled by the interplay of the individual stages. If one of the early steps is slow, its rate determines the overall rate, and the experimental rate law is likely to be simple.

Sometimes the intermediates are relatively stable. It is then easy to determine the concentrations of reactants, intermediates and products at all stages of the reaction. Typical examples of stable intermediates are found in the esterification of di- or poly-carboxylic acids or in the alkylation of aromatic hydrocarbons. Thus, in the alkylation of isopropyl benzene with butene, and with hydrogen fluoride as a catalyst, the reactions are

$$\underset{\text{(A)}}{\text{isopropylbenzene}} \rightarrow \underset{\text{(B)}}{\text{isopropyl-s-butylbenzene}} \rightarrow \underset{\text{(C)}}{\text{isopropyl-di-s-butylbenzene}}$$

and the proportions of A, B and C change as shown in *Fig.* 9*a*.

When these consecutive reactions are each of simple order, and none are reversible, the rates of formation of B and C can be expressed in terms of the velocity constants and can be predicted if the velocity constants and the initial conditions of concentration are

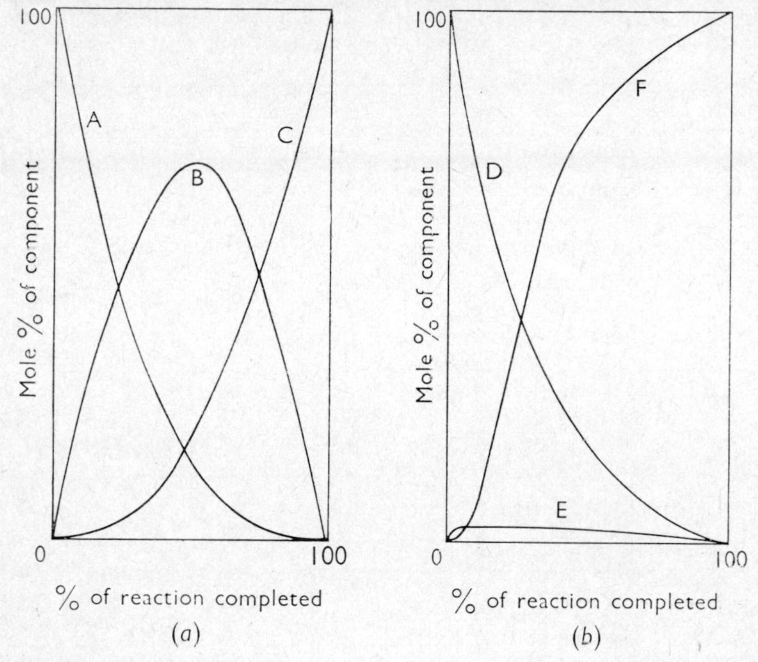

FIG. 9 (*a*) Typical composition/time plots for stable intermediate, *e.g.* A = isopropyl benzene; B = isopropyl-s-butylbenzene; C = isopropyl-di-s-butylbenzene

(*b*) Typical composition/time plots for unstable intermediate, *e.g.* D = diphenylmethyl chloride; E = $(C_6H_5)_2CH^+$; F = Cl^- or $(C_6H_5)_2CHOH$ or H^+

known. The mathematics of this is fully explained in the reference books listed on p. 79, but it is too complicated for discussion here.

It is more common to find that the intermediates are reactive; their concentrations never rise above low values and the rate, $-d[\text{intermediate}]/dt$, is very small compared with other rate terms (*Fig.* 9*b*). Hence in the differential equations which describe the rate of removal or appearance of reactants, intermediates and products, all the differentials $d[\text{intermediate}]/dt$ can be set equal to zero. This is the *principle of stationary states* of reactive intermediates, a most important principle that we shall illustrate in the discussion which follows. Once the principle is assumed for the intermediates in some complex scheme, it is usually possible to express the overall rate as a law predicted from the chosen reaction scheme, and this predicted law can be compared with the experimental law. This is the first step in finding a satisfactory mechanism for a complex reaction. To be acceptable, a proposed mechanism must give the correct *form* of the rate law. Where possible, it should also be tested quantitatively. Progress here has been slow, because the rate constants of the proposed reaction steps cannot be calculated accurately, as explained earlier, and very few of these steps are accessible by direct experiment as they cannot easily be isolated from other reactions. However, the reverse procedure has been fruitful. Rate constants of some individual steps can be determined from the rate of the overall reaction and a rate law predicted from a mechanism which seems well-established (for example, by positive identification of the intermediates). These constants, and those that are accessible directly, can be checked by application to other reactions in which the individual steps occur, thus building up our quantitative knowledge of rate constants.

Some simple examples

We shall first examine some reactions where the experimental rate law is found to be simpler than would be expected from the chemical equation for the reaction.

(1) The hydrolysis of t-butyl chloride might be expected to follow a second-order rate law just like the reaction between methyl bromide and OH^-. However, with low concentrations of water in nearly anhydrous formic acid, it follows a first-order law:

$$(CH_3)_3CCl + H_2O \longrightarrow (CH_3)_3COH + H^+ + Cl^- \qquad (R5)$$

$$-d[(CH_3)_3CCl]/dt = k[(CH_3)_3CCl]$$

This is not a 'pseudo-law', simulated because the concentration of water is high enough to be virtually constant, as it would be in aqueous solution. It appears to be due to the reaction proceeding in two steps, the first heterolytic split of the C–Cl bond being slow and so determining the rate.

$$(CH_3)_3CCl \xrightarrow{1} (CH_3)_3C^+ + Cl^- \qquad \textit{slow}$$

$$(CH_3)_3C^+ + H_2O \xrightarrow{3} (CH_3)_3COH + H^+ \quad \textit{fast}$$

(As we shall see later, step 1 may be reversible with the reverse reaction, 2, slow relative to step 3.) It is, of course, necessary to inquire why this path is faster than the direct bimolecular reaction, and why the first unimolecular step is a heterolytic split rather than a homolytic split to free radicals. There is evidence from other work that the heterolytic split becomes easier with increasing substitution of alkyl groups on the carbon to which the chlorine is attached, because the bond becomes more polarized. In polar solvents of high dielectric constant the ions resulting from heterolytic bond-split can easily be separated and are stabilized by solvation with the polar solvent.

(2) Another interesting set of reactions for which acids or bases act as catalysts includes the halogenation of ketones in aqueous solutions. These are step-by-step substitutions of halogen for hydrogen in alkyl groups, *e.g.*

$$CH_3COCH_3 + I_2 \longrightarrow CH_3COCH_2I + H^+ + I^-$$

The substitution stops here in acid solution, but proceeds much further in solutions of high pH. However, in solutions of low or moderate pH experiment shows that the rate law is

$$-d[I_2]/dt = k[CH_3COCH_3][Y]$$

where Y stands for acid or base present. Strictly the rate law is the sum of products such as that shown, one term for each general acid or base present. It will be seen that the rate is first-order in ketone and in catalyst, but 'zero'-order in iodine. A similar law is found with the other halogens. It is also similar for the catalysed racemization of substituted optically-active ketones. These experimental findings can be rationalized if the reactions have a common first step which is slower than the subsequent different steps. With bases, for example, the ketones probably form a carbanion by loss of a proton in a slow step; with acids, the carbonyl oxygen is protonated and then the enol is formed before the fast step with iodine. Thus with base catalysts:

$$CH_3COCH_3 + B \longrightarrow CH_3COCH_2^- + BH^+ \quad (slow)$$

$$CH_3COCH_2^- + I_2 \longrightarrow CH_3COCH_2I + I^- \quad (fast)$$

$$BH^+ + H_2O \rightleftharpoons B + H_3O^+ \quad (both\ fast)$$

(3) Another reaction for which the rate law is simpler than expected is the oxidation of iodide ion by persulphate ion. In M/10 solutions, this reaction proceeds at a rate which can conveniently be followed by titrating samples of the reacting mixture at intervals to determine the iodine formed.

$$S_2O_8^{2-} + 2I^- \longrightarrow 2SO_4^{2-} + I_2$$

The reaction consists of breaking the O–O bond and making the I–I bond, with the simultaneous transfer of the electrons. It is possible to envisage a transition complex for this reaction proceeding in one stage. However, the rate law shows that the reaction proceeds by a different route. The experimental rate law is

$$-d[S_2O_8^{2-}]/dt = d[I_2]/dt = k[S_2O_8^{2-}][I^-]$$

Thus the transition complex contains one persulphate and one iodide ion. In one suggested route the slow step is

$S_2O_8^{2-} + I^- \rightleftharpoons O_3SO^{2-} \cdot\cdot OSO_3^- \cdot\cdot I \rightarrow SO_4^{2-} + SO_4^- + I$ .. (*slow*)

followed by $SO_4^- + I^- \longrightarrow SO_4^{2-} + I$ .. .. (*fast*)

and $I + I \longrightarrow I_2$ .. .. .. (*fast*)

Unfortunately it is difficult to find positive evidence for the iodine atoms and sulphate ion-radicals postulated as intermediates—or indeed to identify any other intermediates in this reaction.

(4) An apparently similar reaction has a more complicated rate law. This is a reaction examined by Harcourt and Esson in their pioneering work in reaction kinetics:

$$H_2O_2 + 2I^- + 2H^+ \longrightarrow 2H_2O + I_2$$

It can also be followed by titration of the iodine at suitable intervals. By varying the concentrations of the reactants H_2O_2 and I^- in turn, at constant pH, it is found that the rate is proportional to $[H_2O_2]$ and to $[I^-]$. The constant of proportionality varies linearly with the $[H^+]$, however; thus

$$-d[H_2O_2]/dt = k[H_2O_2][I^-]\,(1 + k'[H^+])$$
$$= k[H_2O_2][I^-] + kk'[H_2O_2][I^-][H^+]$$

This equation is interpreted as indicating two parallel reaction paths; in one path the rate is determined by a simple bimolecular reaction:

$H_2O_2 + I^- \longrightarrow$ intermediates .. .. .. (*slow*)

intermediates $+ I^- \xrightarrow{2H^+} 2H_2O + I_2$ .. .. (*fast*)

In the second path, the slow step is *either* a termolecular step

$$H_2O_2 + I^- + H^+ \longrightarrow \text{intermediates}$$

or comes immediately after an initial equilibrium such as

$H^+ + I^- \rightleftharpoons HI$ .. .. .. (*rapidly established*)

$HI + H_2O_2 \longrightarrow$ intermediates .. .. .. (*slow*)

intermediates $+ H^+ + I^- \longrightarrow 2H_2O + I_2$ .. (*fast*)

(5) The next reaction we shall examine is applied in analytical chemistry to obtain bromine solutions of known concentrations. It is chemically similar to the better known reaction which produces iodine.

$$BrO_3^- + 5Br^- + 6H^+ \longrightarrow 3Br_2 + 3H_2O$$
$$IO_3^- + 5I^- + 6H^+ \longrightarrow 3I_2 + 3H_2O$$

However, whereas the second reaction is very rapid in all but the most dilute solutions, the first can conveniently be followed in, say, M/50 solutions by titrating the bromine in samples which have been neutralized with bicarbonate ions after withdrawal from the reaction mixture. By changing the concentrations of each reacting ion in turn, the rate is found to depend upon the law

$$-d[BrO_3^-]/dt = (\tfrac{1}{3})d[Br_2]/dt = k[BrO_3^-][Br^-][H^+]^2$$

This could indicate a single-stage, rate-determining step (i) in which one bromate ion, one bromide ion and two hydrogen ions meet to form a transition complex which then forms intermediates; these react rapidly with further bromide and hydrogen ions to give bromine and water.

$$\text{(i)}\quad 2H^+ + Br^- + BrO_3^- \rightleftharpoons \text{(complex)} \longrightarrow \text{intermediates}$$

However, there are other possibilities. The discussion on the conversion of ammonium cyanate to urea (p. 33) showed that the rate law really defines the stoichiometry of the transition complex of an elementary reaction without defining uniquely the actual species that meet to form the complex. Thus, it is possible for two or more of the ions to be in equilibrium with precursors of the transition complex. It is unlikely that two ions of like charge would form a precursor, so that the likely variations are shown in (ii), (iii) and (iv).

(ii)

$$H^+ + Br^- \rightleftharpoons HBr$$
$$HBr + H^+ + BrO_3^- \rightleftharpoons \text{complex} \longrightarrow \text{intermediates}$$

(iii)

$$H^+ + Br^- \rightleftharpoons HBr$$
$$H^+ + BrO_3^- \rightleftharpoons HBrO_3$$
$$HBr + HBrO_3 \rightleftharpoons \text{complex} \longrightarrow \text{intermediates}$$

(iv)

$$H^+ + BrO_3^- \rightleftharpoons HBrO_3$$
$$HBrO_3 + H^+ \rightleftharpoons H_2BrO_3^+$$
$$H_2BrO_3^+ + Br^- \rightleftharpoons \text{complex} \longrightarrow \text{intermediates}$$

With these kinds of pre-equilibria, in each of which a number of original reactants form a single precursor, it proves impossible to distinguish between the various reaction paths by any kinetic or rate measurements. If one intermediate could be positively identified and the absence of the others proved, say by spectroscopic analysis, one path could be preferred. But it is extremely difficult to prove the absence of an intermediate that has been postulated as reactive and so need be present in only small proportions; only if some of the proposed intermediates can be shown to be unreactive, so demanding high concentrations which could be detected, or if the intermediates are extremely unlikely on, for example, structural or electrostatic grounds, can some selection be made. Thus (iii) is more likely to occur than (iv), as $HBrO_3$ is a very weak base. But there is little direct evidence about the early steps in this reaction. It is, however, known that the most likely intermediates after the rate-determining step are BrO_2^- and BrO^- ions, or the conjugate acids, as these are rapidly attacked by bromide ions in acid solutions to give bromine and water.

A likely scheme is therefore:

$$\text{Prior equilibrium steps}\begin{cases} H^+ + Br^- \underset{-1}{\overset{1}{\rightleftharpoons}} HBr \ (\textit{rapidly established}) \\ H^+ + BrO_3^- \underset{-2}{\overset{2}{\rightleftharpoons}} HBrO_3 \ (\textit{rapidly established}) \end{cases}$$

Rate-determining step:

$$HBr + HBrO_3 \rightleftharpoons \text{transition complex} \xrightarrow{3} HBrO + HBrO_2$$

$$\text{Fast later steps}\begin{cases} HBrO_2 + HBr \longrightarrow 2HBrO \\ HBrO + HBr \longrightarrow H_2O + Br_2 \end{cases}$$

From the first two pairs of balanced reactions, assuming 3 to be slow,

$$[HBr] = K_{1,-1}[H^+][Br^-]$$

$$[HBrO_3] = K_{2,-2}[H^+][BrO_3^-]$$

$$\therefore \frac{d[Br_2]}{dt} = k_3[HBr][HBrO_3] = k_3K_{1,-1}K_{2,-2}[Br^-][BrO_3^-][H^+]^2$$

More complex rate laws

The hydrolysis of diphenylmethyl chloride

$$(C_6H_5)_2CHCl + H_2O \longrightarrow (C_6H_5)_2CHOH + Cl^- + H^+ \quad .. \quad (R6)$$

might be expected to follow a second-order rate law, like reaction R4 (p. 21) or a first-order law like reaction R5 (p. 46). The experimental rate law is more complicated, however, and as it is typical of the rate law for many reactions we shall examine it and the mechanism it indicates at some length. One of our aims is to demonstrate that the experimental rate law and the mechanism of a complex reaction must be treated together.

Experimentally, it was found that the initial rates are proportional to the concentration of diphenylmethyl chloride, but later in the reaction they are slower than would be predicted from the initial rates (*see Fig. 9b*). With the possibility of a reversible reaction in mind, like the reaction $H_2 + I_2 \rightleftharpoons 2HI$, the effect of each of the products on the initial rates was investigated. It was found that the alcohol and the hydrogen ion had no effect, but the chloride ion slowed the initial rate. This finding at once shows that the complete reaction is not reversible, but some reversible reaction of chloride ion is involved. Also, simple semi-quantitative tests showed that, with excess of chloride ion, the rate is inversely proportional to the concentration of chloride ion, and does not follow a law similar to eqn 8 (p. 22) for a reversible elementary reaction. From this evidence, the rate law may be of the form

$$\text{Rate} = \frac{[(C_6H_5)_2CHCl]}{\alpha + \beta[Cl^-]} \qquad \cdots \qquad \cdots \qquad (10)$$

or of the form

$$\text{Rate} = \frac{[(C_6H_5)_2CHCl]}{\alpha'[Cl^-]} + \beta'$$

The second equation could be tested by plotting (Rate) against $[(C_6H_5)_2CHCl]/[Cl^-]$, and the fit was poor. To see how to test the first equation (or any equation like it with a single term in the numerator and the sum of terms in the denominator) invert the equation and rearrange to give

$$\frac{[(C_6H_5)_2CHCl]}{\text{Rate}} = \alpha + \beta[Cl^-]$$

Thus if eqn 10 is obeyed, a plot of $[(C_6H_5)_2CHCl]/\text{Rate}$ against $[Cl^-]$ should be linear, and from the intercept slope α and β can be evaluated.

This plot shows that eqn 10 is followed. Now a likely first step is the heterolytic split of the C–Cl bond. If this is assumed to be reversible we have a scheme which is an extension of that proposed for the hydrolysis of t-butyl chloride (p. 46).

$$(C_6H_5)_2CHCl \underset{2}{\overset{1}{\rightleftharpoons}} (C_6H_5)_2CH^+ + Cl^-$$

$$(C_6H_5)_2CH^+ + H_2O \xrightarrow{3} (C_6H_5)_2CHOH + H^+$$

We shall denote the three reactions by 1, 2 and 3, the corresponding rate constants by k_1, k_2 and k_3, and the radical $(C_6H_5)_2CH$ as A. Treating each reaction as elementary, the rate equations for the species involved are:

$$d[Cl^-]/dt = (-\,d[ACl]/dt) = k_1[ACl] - k_2[A^+][Cl^-]$$

$$d[A^+]/dt = k_1[ACl] - k_2[A^+][Cl^-] - k_3[A^+][H_2O]$$

$$d[AOH]/dt = k_3[A^+][H_2O]$$

Applying the principle of stationary states to the concentration of the reactive intermediate, A^+, $d[A^+]/dt$ is assumed negligible in comparison with other rate terms. The second differential equation then gives

$$[A^+] = \frac{k_1[ACl]}{k_2[Cl^-] + k_3[H_2O]}$$

We may note that if $[A^+] \ll [ACl]$, which is necessary for the assumption of stationary states to be valid, then k_1 must be much smaller than $k_2[Cl^-] + k_3[H_2O]$. As $[Cl^-] \simeq 0$ in the early stages of reaction, $k_3[H_2O]$ must be much greater than k_1. Substituting for $[A^+]$ in the third differential equation, the overall rate is

$$\frac{d[AOH]}{dt} = \frac{k_1 k_3[ACl][H_2O]}{k_2[Cl^-] + k_3[H_2O]} \quad .. \quad .. \quad (11)$$

Also, substituting for $[A^+]$ in the first differential equation, it can easily be shown that $-d[ACl]/dt = d[AOH]/dt$, as the chemical equation demands.

Comparison of eqn 11 with the experimental eqn 10 shows that they can be equated if $[H_2O]$ is assumed constant. This will be true with moderately dilute solutions of the diphenylmethyl chloride in aqueous acetone. Writing $k_3[H_2O]$ as k_3' the comparison then shows that $\alpha = 1/k_1$ and $\beta = k_2/k_1k_3'$. Thus k_1 and the ratio k_2/k_3' can be evaluated.

The terms in the denominator of eqn 11, or the equivalent eqn 10, arise from the reactions 2 and 3, where there is competition for the intermediate, A^+. In general, terms like these arise from competitive reactions, and when they are found in an experimental rate-equation the mechanism must include such competitive steps.

If the competition is greatly in favour of reaction 3, either because k_3 is very large compared with k_2 (an internal property of reactions 2 and 3) or because $[H_2O]$ is much greater than $[Cl^-]$, which can be controlled by the choice of solvent and reaction conditions, then $k_3[H_2O] \gg k_2[Cl^-]$ and eqn 11 reduces to

$$\text{Rate} = k_1[ACl]$$

It is important to notice that this simplification does *not* require that $[H_2O]$ is so large as to be considered constant. The rate is then controlled by the rate of the first reaction. This is the condition visualized in the discussion of the hydrolysis of t-butyl chloride, but it can now be seen that the more general case of a reversible first step can include the simple law as an extreme case.

At the other extreme, when $k_2[Cl^-] \gg k_3[H_2O]$, the rate law would be

$$\text{Rate} = \frac{k_1 k_3[ACl][H_2O]}{k_2[Cl^-]} \quad .. \quad .. \quad .. \quad (12)$$

This rate law would hold if high concentrations of chloride were added at the start, or, perhaps, late in a normal run.

It is important to realize where equilibria 1 and 2 differ from the equilibria discussed on p. 48 under the reaction between bromide, bromate and hydrogen ions. There, the equilibria were 'association' reactions, and the forward step produces one species to react in the rate-determining step. Here two species are produced, and although both can affect the reverse step of the equilibrium, only one species (the ion A^+) affects the rate of reaction 3.

It is also important to note that the overall rate can equal, but never exceed, the rate of the initial reaction, 1. For at one extreme, when $k_2[Cl^-] \gg k_3[H_2O]$, the rate is

$$k_1[ACl] \cdot \frac{k_3[H_2O]}{k_2[Cl^-]}$$

i.e. (rate of reaction 1) × (factor $\ll 1$),

and as we move to the other extreme the rate approaches that of reaction 1.

It is not possible to assume that d[intermediate]/dt can be set equal to zero for some reactions. This occurs if the concentration of intermediate becomes comparable with those of the reactants (or products at later stages). It is not easy to solve the differential equations for these cases. However, if the rate constants for steps 1, 2 and 3 are known, or can be estimated, the variation of each concentration with time can be derived on digital or analogue computers and compared with the experimental concentration/time relationships. If the match is not good, the velocity constants can be changed (this is particularly easy with an analogue computer) and the values found which give the best fit. This is proving an extremely useful technique.

Many different types of reaction follow mechanisms similar to the three-stage scheme found in the hydrolysis of diphenylmethyl chloride. Examples include the nitration of aromatic hydrocarbons, electron-transfer reactions such as $2Fe^{2+} + Tl^{3+} \longrightarrow 2Fe^{3+} + Tl^+$, the reduction of dichromates or cupric salts by dissolved hydrogen, and many reactions catalysed by acids or bases. We shall not examine these in any detail, but turn instead to four general types of reaction where the basic three-stage pattern gives a good account of the kinetics of the reactions. The basic pattern is

$$\text{Reactants } (+\text{ catalyst or solvent}) \underset{2}{\overset{1}{\rightleftharpoons}} \text{intermediates}$$

$$\text{Reactants} + \text{intermediates} \xrightarrow{3} \text{products } (+\text{ catalyst if present})$$

Of course there are many elaborations and extensions of this simple scheme—for example additional equilibria can be found before the rate-determining step—but these merely involve extension of the basic pattern.

Thermal decomposition of gases

There are many thermal decompositions which show a first-order rate law when carried out in solution or at high pressures. These first-order decompositions in the gas phase appeared rather paradoxical when the collision theory of gas reactions was applied to them. If a molecule decomposes on colliding with another, provided that they share enough kinetic energy, all rates of gaseous reactions should depend on the frequency of collisions and hence on the product of two concentration terms, and so the rate law should always be second-order. The resolution of this paradox by Lindemann is interesting in itself, but in addition it provides most important evidence for the idea of 'activation'. The decompositions are now regarded as complex reactions rather than elementary, and the complete reaction consists of several stages. The molecule must first be activated by exchange of energy in collisions. The molecule with sufficient energy does not decompose immediately, however, perhaps because the energy gained from collision must be transferred through some other vibrations into vibrational energy in the bond which is to break. There is time for deactivation by loss of energy through other collisions. The reaction scheme is therefore (A* represents an energized molecule A)

$$A + A \xrightarrow{k_1} A^* + A$$

$$A^* + A \xrightarrow{k_2} A + A$$

$$A^* \xrightarrow{k_3} \text{decomposes}$$

The principle of stationary states (p. 45) can be applied to the concentration of A*, the active intermediate. Thus

$$\frac{d[A^*]}{dt} = k_1[A]^2 - k_2[A][A^*] - k_3[A^*] = 0$$

$$\therefore [A^*] = \frac{k_1[A]^2}{k_2[A] + k_3}$$

$$\therefore \text{Rate of decomposition} = k_3[A^*] = \frac{k_1 k_3 [A]^2}{k_2[A] + k_3}$$

Any such decomposition is therefore of variable order, the order changing as [A] is increased by increasing the pressure. If the pressure is made low enough, $k_2[A]$ can be made very much lower than k_3, while if it is made high enough, $k_2[A]$ can be much larger than k_3. Thus at the extremes:

$$[A] \text{ very low:} \quad \text{Rate of decomposition} = k_1[A]^2$$

$$[A] \text{ very high:} \quad \text{Rate of decomposition} = k_1k_3[A]/k_2$$

Hence the decomposition follows a first-order law at high pressures and a second-order law at low pressures. Experimentally, it is found that the first-order law is maintained at lower pressures with more complex molecules. The rate of decomposition of simple molecules, for example the dissociation of bromine molecules,

$$Br_2 + M \longrightarrow 2Br + M$$

can be measured by special techniques such as shock-tube studies (*see* p. 4). These show that the rate is proportional to the pressure of the bromine (which is usually some 100 Pa) and the pressure of the carrier gas, often argon or nitrogen, even when the total pressure is several hundred kPa. On the other hand, it has been known for a long time that with decompositions of large molecules, such as cyclobutane,

$$\begin{array}{l} CH_2\text{—}CH_2 \\ \,|\qquad\;\; | \\ CH_2\text{—}CH_2 \end{array} \longrightarrow 2C_2H_4$$

the first-order law holds down to a few kPa. All these differences have been shown to be in agreement with the theory of activation by exchange of energy, and the change from first-order to second-order rate laws is one of the most important pieces of evidence for the soundness of the theory.

It is customary to describe these decompositions as unimolecular under all experimental conditions, irrespective of the rate law that holds. This is because bond-breaking and bond-making occur in only one molecule in forming the transition complex. However, this custom is only satisfactory because we do not regard the change of energy as occurring through a transition complex; at low pressures, the rate-determining step is certainly a binary collision even if the event is not usually called a bimolecular reaction.

In solution, of course, these problems do not arise and the rate is first-order at all concentrations, because even in dilute solutions the energization is amply maintained by collision with solvent molecules.

The decomposition of N_2O_5 deserves special mention. It has often been erroneously described as an elementary unimolecular first-order reaction. However, the *molecular* equation for the decomposition is

$$2N_2O_5 \longrightarrow 2N_2O_4 + O_2$$

and it is clear that a transition complex could not be formed from one molecule of N_2O_5 and immediately yield these products. Moreover, the reaction is first-order at pressures much lower than 100 Pa, unlike normal gas-phase decompositions of relatively simple molecules, and the rate is greatly increased by adding nitric oxide.

These effects have been interpreted by a reaction scheme devised by Ogg:

$$N_2O_5 \underset{2}{\overset{1}{\rightleftharpoons}} NO_3 + NO_2 \qquad \textit{(rapidly established)}$$

$$NO_3 + NO_2 \xrightarrow{3} NO_2 + NO + O_2 \qquad \textit{(slow step)}$$

$$NO_3 + NO \xrightarrow{4} 2NO_2 \qquad \textit{(fast)}$$

In the absence of added NO, step 3 controls the rate; as soon as NO is formed it is removed by step 4. The rate of reaction is then found by setting the expression for $d[NO_3]/dt$ equal to zero, giving an expression for $[NO_3]$. When substituted in the rate equation for 3 there results

$$\frac{d[O_2]}{dt} = \frac{k_1k_3[N_2O_5]}{k_2 + k_3} \simeq \frac{k_1k_3}{k_2}[N_2O_5]$$

Thus the experimental rate constant is equal to $k_1/k_2 \times k_3$, *i.e.* to (an equilibrium constant) × (a second-order rate constant). There is no reason why this should be affected by any changes of total pressure.

In the presence of added NO, step 3 is by-passed and the rate is controlled entirely by step 1. Hence

$$\left(\frac{d[O_2]}{dt}\right)_{NO} = k_1[N_2O_5]$$

Step 1 is a truly unimolecular reaction and it does show the expected increase in order at pressures below 100 kPa.

Reaction between ions in aqueous solution

It is sometimes thought that reactions between ions are extremely fast. This is true for precipitation and neutralization reactions. It has proved extremely difficult to measure the velocity of reactions such as

$$H_3O^+ + OH^- \longrightarrow 2H_2O$$

$$\text{or}\quad CO_2 + OH^- \longrightarrow HCO_3^-$$

Special techniques have recently been evolved, however, which allow these fast rates to be measured, at least to orders of magnitude. It appears that these reactions occur almost as rapidly as the reacting ions can diffuse together. The rate is not limited by the chemical structure of the reactants but by the diffusion coefficient of the ions through the solvent medium. On this basis, theories of diffusion show that the maximum velocity constant for a bimolecular reaction in water is about 10^{10} l mol^{-1} s^{-1}, and so with molar solutions the half-life (p. 35) is about 10^{-10} s. This rate clearly necessitates very special experimental techniques.*

However, for many ionic reactions the rate constant is much smaller. Examples of reactions that can easily be followed by ordinary titration methods are:

$$2Fe^{3+} + 2I^- \longrightarrow 2Fe^{2+} + I_2$$

$$S_2O_8^{2-} + 2I^- \longrightarrow 2SO_4^{2-} + I_2$$

$$H_2O_2 + 2I^- + 2H^+ \longrightarrow 2H_2O + I_2$$

$$BrO_3^- + 5Br^- + 6H^+ \longrightarrow 3H_2O + 3Br_2$$

In addition, the exchange of ligands in some complex ions can be followed by colour changes.

Investigations of the rates of these reactions showed that they behaved like reactions between molecules, and that the rate laws indicated that most of the reactions were complex and consisted of several stages of elementary reactions between ions. Where the rates of the elementary reactions could be measured, a curious feature emerged. The velocity constants of bimolecular, second-order reactions change regularly with the addition of 'neutral' salts that do not affect the reaction chemically, but alter the density of ionic charges in the solvent medium. The alteration of charge density was known to alter the activity of any ionic solute and, in solutions of low and moderate concentration, increase of ionic density decreases the activity. Hence, if rates are proportional to activities and not to concentrations, all rates would be decreased when neutral salts were added. However, this only happened if the reaction was between ions of *opposite* sign, such as the ferric–iodide reaction. When the reactants were of *like* sign, as in the persulphate–iodide reaction, the rate actually increased upon addition of neutral salts.

The solution of this problem was offered in 1928 by Brønsted and Bjerrum. They proposed that the ions are in equilibrium with an

---

* *See* E. F. Caldin, *Fast Reactions in Solution*, Oxford: Butterworths Scientific Publications, 1964.

'ionic complex' which slowly leaked to products.* Thus with ions A of charge z_A, and B of charge z_B,

$$A^{z_A} + B^{z_B} \underset{2}{\overset{1}{\rightleftharpoons}} X^{(z_A + z_B)}$$

$$X^{(z_A + z_B)} \xrightarrow{3} \text{products}$$

The last reaction is regarded as so slow relative to 1 and 2 that the first two are in equilibrium. However, this equilibrium is properly expressed in terms of activities, $a = \gamma c$, where γ is the activity coefficient and c the concentration. Then (omitting charges for clarity)

$$K_a = \frac{a_X}{a_A\, a_B} = \frac{\gamma_X\,[X]}{\gamma_A\,\gamma_B[A][B]}$$

$$\therefore \frac{d[\text{products}]}{dt} = k_3[X] = k_3K_a\,\frac{\gamma_A\gamma_B}{\gamma_X}\,[A][B]$$

Hence if $$\frac{d[\text{products}]}{dt} = k_{exp}\,[A][B],$$

$$k_{exp} = k_3K_a\,\frac{\gamma_A\gamma_B}{\gamma_X}$$

Now the theory of Debye and Hückel relates γ for an ion to the charge on the ion and the 'ionic strength', I, of the medium. The value of I is a measure of the ionic charge-density, and it is defined as $\frac{1}{2}\sum_i m_i z_i^2$; here $m_i \approx c_i$. The relation for aqueous solutions at 25 °C is

$$\log_{10}\gamma_i = -\tfrac{1}{2}\,z_i^2\,\sqrt{I}$$

$$\begin{aligned}\text{Then } \log_{10} k_{exp} &= \log_{10} k_3K_a + \log_{10}\gamma_A + \log_{10}\gamma_B - \log_{10}\gamma_X \\ &= \log_{10} k_3K_a - \tfrac{1}{2}z_A^2\sqrt{I} - \tfrac{1}{2}z_B^2\sqrt{I} + \tfrac{1}{2}(z_A + z_B)^2\sqrt{I} \\ &= \log_{10} k_3K_a + z_Az_B\sqrt{I}\end{aligned}$$

For two ions of *like charge*, z_Az_B must be positive, and k_{exp} should increase as I is made larger. For ions of *unlike charge*, z_Az_B will be negative, and k_{exp} should decrease as I is made larger.

Not only is the equation in agreement qualitatively with the experimental results, but it was found to agree quantitatively at low ionic strengths.

Some deviations have since been found with particular ions and neutral salts, but the general picture remains established. It is

* We would now regard this complex as a transition complex; the older treatment is very close to the transition-state treatment for non-ideal reactants (*see* Laidler; reference books, p. 79).

striking evidence for the existence of the ionic intermediate and for the concept of the three-stage reaction. It is unlikely that effects like these would show up in gas reactions as the deviations from ideal behaviour are so much less pronounced.

Enzyme-catalysed reactions

The basic mechanism can be applied to reactions catalysed by an enzyme, E. If there is one reactant or 'substrate', S, it forms a complex reversibly with the enzyme, and the complex slowly 'leaks' to products, regenerating the enzyme

$$E + S \underset{2}{\overset{1}{\rightleftharpoons}} ES$$

$$ES \xrightarrow{3} P + E$$

This reaction scheme was treated by Michaelis and Menton, and later by Briggs and Haldane. In following the kinetics of enzyme reactions it is often convenient to measure the initial rates of different reaction mixtures rather than to follow each run through. If the initial concentrations taken are $[S]_0$ and $[E]_0$, then as soon as a concentration [ES] of complex is formed

$$[S] = [S]_0 - [ES]$$

and

$$[E] = [E]_0 - [ES].$$

But $[E]_0$ and hence [ES] are usually $\ll [S]_0$. Hence $[S] \simeq [S]_0$. If these concentrations are used in the stationary-state treatment for the concentration of complex,

$$\frac{d[ES]}{dt} = 0 = k_1[S]_0\{[E]_0 - [ES]\} - (k_2 + k_3)[ES]$$

$$\therefore \text{Rate} = k_3[ES] = \frac{k_3k_1[E]_0[S]_0}{k_2 + k_3 + k_1[S]_0}$$

Thus with $[S]_0$ constant the initial rate is proportional to $[E]_0$, but when $[S]_0$ is increased with $[E]_0$ constant, the rate reaches a limiting (maximum) value equal to $k_3[E]_0$. Physically this occurs when the enzyme is fully saturated because the balance in reactions 1 and 2 is pushed far to the right by the increase in $[S]_0$.

Reactions catalysed by surfaces

Many surface-catalysed reactions of gases occur by the reactants becoming chemisorbed on sites on the surface to give 'surface-compounds'. These are often fragments of the reactants bonded to the surface by covalent or ionic bonds. When these bonds are not too strong, the fragments can react to give first the chemisorbed product and then the gaseous products.

When the catalyst is very porous and the catalysed reaction very rapid, the rate may be controlled by slow diffusion of the reactants to the surface. The products of the reaction may also hinder this diffusion.

Even when the surface is readily accessible, the chemisorption itself may be slow. This occurs, for example, in the synthesis of ammonia from hydrogen and nitrogen, over iron catalysts, when the slow step is the chemisorption of the nitrogen as atoms. In the reverse reaction, the decomposition of ammonia to hydrogen and nitrogen, the desorption of the nitrogen atoms from the surface to give nitrogen molecules determines the rate.

There are other catalysed reactions in which the chemisorption–desorption processes are so fast that the surface fragments are virtually in equilibrium with the gas-phase reactants, evaporating from the occupied surface at the same rate as the reactants are chemisorbed on vacant sites. Thus:

$$\text{Reactants} + \text{surface sites} \underset{2}{\overset{1}{\rightleftharpoons}} \text{surface compounds} \quad (\textit{rapidly established})$$

$$\text{Surface compounds} \xrightarrow[3]{\text{slow}} \text{surface products} \quad (\textit{slow})$$

$$\text{Surface products} \rightleftharpoons \text{gaseous products} + \text{free surface sites} \quad (\textit{rapid})$$

This scheme was developed by Langmuir and Hinshelwood and has been applied, with variation, to many catalysed reactions. We shall illustrate it by reference to the decomposition of phosphine over tungsten or molybdenum surfaces. Suppose there are N_s surface sites available for chemisorption on the surface of the catalyst. If the pressure of phosphine is p, and the fraction θ of available surface sites is occupied by surface compounds, then a stationary-state treatment of the steps 1, 2 and 3 gives

$$\frac{\text{d[surface compounds]}}{\text{d}t} = 0 = k_1 p.N_s(1 - \theta) - (k_2 + k_3).N_s\theta$$

$$\therefore \frac{\theta}{1 - \theta} = \frac{k_1 p}{k_2 + k_3}$$

$$\text{or} \qquad \theta = \frac{k_1 p}{k_2 + k_3 + k_1 p}$$

$$\therefore \frac{\text{d[products]}}{\text{d}t} = k_3 N_s \theta = \frac{k_1 k_3 N_s p}{k_2 + k_3 + k_1 p} \qquad (13)$$

This equation is formally similar to the Briggs–Haldane equation for enzyme reactions. Like that equation, it predicts 'saturation' of the surface sites with high values of p, and a rate which is then independent of p (being equal to k_3N_s, in comparison with $k_3[E]_0$ for the enzyme reaction) so giving a 'zero-order' rate law. Experimentally this holds for pressures above 100 Pa of phosphine over molybdenum, and at rather lower pressures over tungsten. At low pressures, the rate is proportional to the phosphine pressure, so giving a first-order law; this is found at pressures between 1 and 10^{-1} Pa.

The eqn 13 indicates that the rate will be proportional to the number of active sites, N_s, available on the metal surface. In turn, this is proportional to the area of a catalyst of uniform surface. This has been found true for a variety of solid catalysts and reactions, from simple hydrogenations over metals to 'cracking' reactions (decompositions, rearrangements and dehydrogenations) over insulator oxides like alumina. Faster rates can be obtained by increasing the surface area of catalysts. It is much harder to alter the number of active sites per unit area, and the 'activity' of each site for chemisorption of reactant (*i.e.* k_1/k_2) or reaction of the surface compound (k_3). Attempts have been made to relate these activities to various properties of the catalyst, such as the atomic geometry of the surface, the electronic structure of the bulk metal or of its atoms, or the heats of chemisorption of the reactants on the surface. No simple, common pattern of activity has emerged, although there have been many promising suggestions, except for the generalization that the chemisorbed fragments must be bound fairly strongly to the surface in order to strain or fragment the bonds in the original reactants, but they must not be held so strongly that the chemisorbed fragments require high activation energies for step 3. It is an experimental fact that heats of chemisorption are high when the coverage is low, and fall as the coverage becomes greater. The bonds between reactants and the sites first covered are stronger—and may be too strong for ready reaction in step 3. The chemisorption on sites occupied later is weaker and the fragments can react more readily in 3. Thus all the sites on the surface may be active in chemisorption, but only a small fraction, corresponding to certain heats of chemisorption, may be active in catalysing the reaction. Much effort has been put into identifying these sites in the hope of improving catalysts by increasing the proportion of the active sites, but it may well be that the 'activity' is an inherent property of the chemisorption process and not of the catalyst alone.*

Order and activation energy of a complex reaction

Inspection of the typical rate laws shows that they cannot easily be classed as belonging to particular orders of reaction. The order

* G. C. Bond, *Principles of Catalysis*, Revised second edition. London: The Chemical Society, 1972.

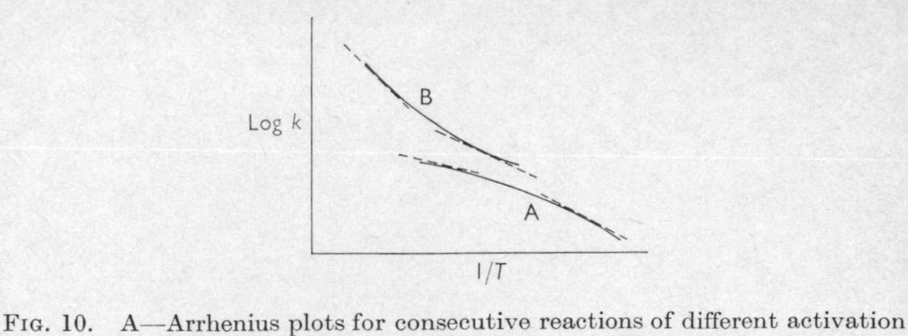

FIG. 10. A—Arrhenius plots for consecutive reactions of different activation energy. B—Arrhenius plots for parallel reactions of different activation energy

varies with the reaction conditions and during a run. It is a mistake to force a classification which is suitable for rate laws for elementary reactions on to complex reactions.

It is doubtful whether a single activation energy, E, and pre-exponential factor, A, can be quoted for each complex reaction. It is now well established that the simple Arrhenius law does not hold for some complex reactions. There are some extreme deviations found with enzyme-catalysed reactions (*Fig.* 4*c*) and in the oxidation of hydrocarbons (*Fig.* 4*d*). The velocity constant at first increases as this temperature is raised, but then decreases. This is attributed to the breakdown of some intermediate by a reaction which does not help the main reaction, *e.g.* by thermal breakdown of the enzyme in 4*c* or the destruction of an intermediate peroxide radical $RO_2\cdot$ in 4*d*. There are some milder but definite deviations which indicate the multi-stage nature of the reaction being examined. Thus with two consecutive stages, as in the hydrolysis of t-butyl chloride, step 1 may be slow at low temperatures and control the rate, but have a high activation energy. Step 3, with a low activation energy, increases more slowly than step 1, and at a higher temperature may limit the rate. This is one explanation of the fact that the Arrhenius plot A in *Fig.* 10 shows a high activation energy (steep slope) at low temperatures and a low activation energy (a lower slope) at high temperatures.

$$(CH_3)_3CCl \xrightarrow{1} (CH_3)_3C^+ + Cl^-$$

$$(CH_3)_3C^+ + H_2O \xrightarrow{3} (CH_3)_3COH + H^+$$

Another explanation connects the change with a temperature-dependence of the A factor, which is ascribed to changes in the number of solvent molecules oriented around the reactants and the transition complex.

On the other hand, with two parallel reactions, the greater part of the reaction will proceed through the low activation-energy path at low temperatures but switch to the high path at high temperatures. This effect, illustrated at B in *Fig.* 10, has been found in the decomposition of nitrosyl chloride where the molecular path

$$2NOCl \longrightarrow 2NO + Cl_2 \qquad (E = 105 \text{ kJ})$$

predominates at low temperatures, but the 'atomic path'

$$NOCl + M \longrightarrow NO + Cl + M \qquad (E = 151 \text{ kJ})$$

$$NOCl + Cl \longrightarrow NO + Cl_2 \qquad (\textit{fast})$$

predominates at higher temperatures.

Rates and equilibrium in reactions that proceed by many stages

We have examined (pp. 22, 27) the relationship between velocity constants and equilibrium constants for reversible elementary reactions

$$\text{reactants} \underset{k_b}{\overset{k_f}{\rightleftharpoons}} \text{products}$$

for which it is possible to write

$$\frac{k_f}{k_b} = K_c$$

because the simple rate laws hold for each reaction.

This very simple relationship holds only for reversible *elementary* reactions. With a complex reaction mechanism, the rate law often—in fact, usually—changes as the reaction proceeds. Consider the iodination of acetone in aqueous acid solution

$$CH_3{\cdot}CO{\cdot}CH_3 + I_2 \xrightarrow{HA} CH_3{\cdot}CO{\cdot}CH_2I + H^+ + I^-$$

This reaction is reversible, as can easily be demonstrated with suitable mixtures, and to a good approximation the equilibrium relationship

$$\frac{[CH_3{\cdot}CO{\cdot}CH_2I]_e[I^-]_e[H^+]_e}{[CH_3{\cdot}CO{\cdot}CH_3]_e[I_2]_e} = K_c$$

must hold, relatively small deviations being due to non-ideal behaviour of the chemical species present. However, this by no means indicates that the actual reactions occurring are the elementary reactions

$$CH_3{\cdot}CO{\cdot}CH_3 + I_2 \longrightarrow CH_3{\cdot}CO{\cdot}CH_2I + H^+ + I^- \quad .. \quad (\alpha)$$

and $$CH_3{\cdot}CO{\cdot}CH_2I + H^+ + I^- \longrightarrow CH_3{\cdot}CO{\cdot}CH_3 + I_2 \quad .. \quad (\beta)$$

As explained earlier, when starting from acetone, iodine and acid catalyst the rate is independent of iodine concentration and proportional to acetone and acid concentrations. The slow step is the formation of the enol form of the acetone. This reaction is clearly reversible, as relatively little of the enol form is produced in the absence of substances which react with it. If iodine is present, it reacts rapidly with the enol form, but as equilibrium is approached, the relative rates of the first and second stages will change, and the enol–iodine reaction may also be reversed. Thus at equilibrium the dynamic balance is maintained by a number of reactions. Only by investigation under conditions far from equilibrium can the actual reactions be identified; they are certainly not the elementary steps α and β.

There is an important principle, called the principle of microscopic reversibility, which states that at equilibrium each and every process is exactly balanced by the reverse process. Consider the reactions discussed for the decomposition of nitrosyl chloride. When this decomposition has reached equilibrium, each of the steps must be balanced by its reverse reaction

Single-step $2NOCl \rightleftharpoons 2NO + Cl_2$

Two-step $\begin{cases} NOCl + M \rightleftharpoons NO + Cl + M \\ Cl + NOCl \rightleftharpoons NO + Cl_2 \end{cases}$

(the reactions $M + Cl_2 \rightleftharpoons 2Cl + M$ must also be in balance). A cycle whereby the forward reaction proceeds entirely by the two-step mechanism and the balance is maintained entirely by the reverse of the molecular step is not allowed because it violates the principle. However, it is very likely that the *effective* paths towards equilibrium are the molecular step and its reverse at low temperatures and the two-step path and its reverse at high temperatures, because the second path has a higher activation energy.

This is a very good example of the possibility of more than one path for a reaction. At equilibrium, thermodynamical arguments assert that each path must be in balance. The proportions of reactions following each path towards equilibrium are not fixed by thermodynamics, but by the energetics of the individual reactions that form the paths.

Thus it can be asserted that for each conceivable elementary reaction a relationship of the form $k_f/k_b = K_c$ must hold, both when the elementary reactions are isolated and when they are part of a complex reaction scheme. It is quite wrong, however, to assume that the existence of these relationships allows us to assert anything about the rate laws for the reactions that actually occur when the system is far from equilibrium, or to identify the actual reactions which predominate in the system at equilibrium.

4. Chain Reactions

We have so far considered elementary reactions completed in one stage, and some complex mechanisms involving intermediates which are formed in one elementary reaction and destroyed in leading to the product. We have already mentioned that the decomposition of dimethyl ether proceeds through a different type of complex reaction, in which the intermediate is regenerated while forming a product molecule. The newly-formed intermediate can react again to give a product, and the successive reactions are 'linked' by the reactive intermediates to form a *chain reaction.*

Chain reactions have some resemblances to the mechanisms discussed for complex reactions, because the reaction scheme often looks like a modification of the three-stage scheme. They differ, however, in two very important respects. The first is that no one step determines the rate, as will be shown later. The second is a consequence of the first. Whereas the rates of the three-stage mechanism can never exceed the rate of the initial step, the rate of the chain reaction may be many orders of magnitude greater than the rate of the initiating step. This property appears particularly clearly in reactions initiated photochemically. In the photochemical reaction between hydrogen and chlorine, for example, the *quantum yield*, which is the number of molecules of product (HCl) formed for each quantum of light-energy absorbed, can be as high as 10^6. In other chain reactions the factor (rate of reaction/rate of initiation), known as the *chain length* and denoted by ν, can also be very high under certain conditions. High values have been observed for other chlorination reactions, for oxidations by dissolved oxygen in solution, and also when high polymers are formed in polymerization reactions. The value of ν is very sensitive, however, to reaction conditions. It can vary with the rate of initiation and with the surface/volume ratio of the reaction vessel, and it can be lowered drastically by the presence of substances called '*inhibitors*' which remove chain centres. Inhibitors can be so effective that, even when their concentration is low, they remove nearly all chain centres before the centres can react in propagating steps. The main reaction is almost completely suppressed. However, the inhibitor is often destroyed by any chain centres formed, and the rate of the main reaction becomes detectable after an *induction period.* Induction periods due to the presence and slow destruction of inhibitors have been observed in chlorination, oxidation and polymerization reactions; this behaviour with inhibitors can be used to distinguish chain from non-chain processes. Of course, heterogeneous and enzyme-catalysed reactions can also be suppressed by poisoning the catalyst, and it is necessary to ensure that a suspected inhibitor of a chain reaction is not a poison for an unsuspected catalyst. There is another cause of induction periods in chain reactions. This is the possibility of *branching reactions* in which more than one centre appears after the consumption of one centre. Classical examples of this occur in oxidation reactions because of the diradical nature of oxygen. In the reaction of hydrogen with oxygen the propagation step is

$$OH + H_2 \longrightarrow H_2O + H$$

and direct branching can occur at high temperatures by

$$H + O_2 \longrightarrow HO + O$$

and

$$O + H_2 \longrightarrow HO + H$$

In these branching reactions additional chain centres are formed in each elementary process. A different process of branching can occur during the oxidation of hydrocarbons. At low temperatures, peroxidic products are formed by reactions such as:

$$RH + O_2 \longrightarrow R + HO_2$$

$$R + O_2 \longrightarrow RO_2$$

$$RO_2 + RH \longrightarrow RO_2H + R$$

The HO_2 radical is not very reactive. However, some hydroperoxide molecules dissociate to give more chain centres in a separate, slow reaction:

$$RO_2H \longrightarrow RO + OH$$

Similar branching can occur through the oxidation or dissociation of fairly stable aldehydes or of H_2O_2. Such branching, occurring long after the original propagation step, is called *degenerate branching*.

Direct and degenerate branching both increase the concentration of chain centres, [X], at rates proportional to the concentration of centres, that is, $(d[X]/dt)_{branching} = k_{branching}[X][reactant]$. In turn, the overall rate of reaction is increased. With direct branching the effect is shown very rapidly after mixing the reactants, and very special techniques are necessary to study the effects. With degenerate branching, however, the centre concentration (and hence the overall rate) may increase slowly over periods of minutes or hours, and so show effects very like those caused by the slow removal of inhibitors. A slow increase in rate from the initial value is never found with elementary reactions. It can be found with some three-stage reactions (cf. the curve for C in Fig. 9*a*). It can also be found when a reaction produces a catalyst among the products (as with some acid–base catalysts); degenerate branching can be regarded as an *autocatalytic process*.

Thus chain reactions show certain characteristics of rate behaviour which help to distinguish them from other complex reactions; however, not all these features are unique to chain reactions, nor does any one chain reaction necessarily show them all:

(*a*) high quantum yields or chain lengths;
(*b*) high sensitivity to inhibitors and to the wall of the vessel;
(*c*) long induction periods due to effective inhibitors;
(*d*) rates which increase from the initial value;
(*e*) long induction periods due to degenerate branching.

We shall not attempt to discuss these effects in detail, but to concentrate on the relation between the overall rate and the rates of the initiation, propagation and termination steps, and to give reasons why chain mechanisms seem to outweigh molecular or three-stage processes in some reactions.

We shall first examine the rate equation for the photochemical reaction between hydrogen and chlorine. This reaction was the first chain reaction to be identified. Nernst and Bodenstein proposed that chlorine atoms are formed by absorption of light of suitable wave-length by chlorine molecules, and these atoms generate a hydrogen atom as well as a product molecule. The hydrogen atom in turn regenerates a chlorine atom, and so on.

$$h\nu + Cl_2 \xrightarrow{1} 2Cl \qquad \textit{initiation step}$$

$$\left.\begin{array}{l} Cl + H_2 \xrightarrow{2} HCl + H \\ H + Cl_2 \xrightarrow{3} HCl + Cl \end{array}\right\} \textit{propagation steps}$$

These two propagation reactions 'shuttle' until one or other centre is removed either by being chemisorbed on the walls, by reaction with inhibitor molecules (In) or by recombination with other atoms. Detailed studies have shown that termination occurs much more often by removal of chlorine than of hydrogen atoms:

$$\left.\begin{array}{l} Cl + \text{wall} \xrightarrow{4} \text{removal} \\ Cl + In \xrightarrow{5} \text{removal} \end{array}\right\} \textit{linear termination}$$

$$Cl + Cl + M \xrightarrow{6} Cl_2 + M \quad \textit{mutual or quadratic termination}$$

To simplify the treatment we shall consider two schemes: in the first, (*a*), termination occurs only by 5, in the second, (*b*), only by 6

As it happens, experimental conditions can often be arranged to ensure that 5 or 6 predominates. The intermediate situations are mathematically complicated and it is then also difficult to establish the rate law experimentally.

Let θ be the number of effective quanta absorbed in the initiation step per unit time per unit volume. Once a steady state is established, application of the principle of stationary states gives:

For case (a)

$$\frac{d[Cl]}{dt} = 2\theta - k_2[Cl][H_2] + k_3[H][Cl_2] - k_5[Cl][In] = 0$$

$$\frac{d[H]}{dt} = k_2[Cl][H_2] - k_3[H][Cl_2] = 0$$

$$\frac{d[HCl]}{dt} = k_2[Cl][H_2] + k_3[H][Cl_2] = 2k_2[Cl][H_2] \quad \left(\text{from } \frac{d[H]}{dt} = 0\right)$$

From $\frac{d[H]}{dt} = 0$ and $\frac{d[Cl]}{dt} = 0$, $[Cl] = \frac{2\theta}{k_5[In]}$

$$\therefore \frac{d[HCl]}{dt} = \frac{4k_2.\theta.[H_2]}{k_5[In]}$$

It is important to notice that the rate of reaction is proportional to θ and inversely proportional to [In] when the termination is linear.

Now [Cl], though small, is not zero.

$$\therefore \frac{d[HCl]}{dt} = \frac{2k_2[Cl][H_2].2\theta}{k_5[Cl][In]}$$

$$\therefore \text{Rate} = \frac{(\text{rate of propagation})\ (\text{rate of initiation})}{(\text{rate of termination})} \quad .. \quad (14)$$

For case (b)

$$\frac{d[Cl]}{dt} = 2\theta - k_2[Cl][H_2] + k_3[H][Cl_2] - 2k_6[Cl]^2[M] = 0$$

$$\frac{d[H]}{dt} = k_2[Cl][H_2] - k_3[H][Cl_2] = 0$$

$$\frac{d[HCl]}{dt} = k_2[Cl][H_2] + k_3[H][Cl_2] = 2k_2[Cl][H_2] \quad (\text{as before})$$

From $\frac{d[H]}{dt} = 0$ and $\frac{d[Cl]}{dt} = 0$, $[Cl] = \sqrt{\frac{\theta}{k_6[M]}}$

$$\therefore \frac{d[HCl]}{dt} = 2k_2[H_2]\sqrt{\frac{\theta}{k_6[M]}}$$

The rate is now proportional to $\sqrt{\theta}$ and inversely proportional to $\sqrt{[\mathrm{M}]}$. Strictly speaking there is a term $k_{6,\mathrm{M}}[\mathrm{M}]$ for each molecular species present.

Again, as $[\mathrm{Cl}] \neq 0$, $\dfrac{\mathrm{d}[\mathrm{HCl}]}{\mathrm{d}t} = 2k_2[\mathrm{H_2}][\mathrm{Cl}]\sqrt{\dfrac{\theta}{k_6[\mathrm{M}][\mathrm{Cl}]^2}}$

$$\therefore \text{Rate} = (\text{rate of propagation})\sqrt{\frac{(\text{rate of initiation})}{(\text{rate of termination})}} \qquad (15)$$

The eqns 14 and 15 for the rate are quite general. They indicate that the fast rate can be achieved by a fast rate of initiation, a fast rate of propagation or a slow rate of termination. No single step determines the rates, but the nature of the steps imposes certain requirements for a chain mechanism to be effective relative to a possible molecular mechanism. In the absence of catalysts or photochemical initiation any thermal initiation is likely to involve breaking bonds, with the probability of a high activation energy, and so be slow. It is essential for the propagation step to be fast relative to the termination reactions, and a chain mechanism cannot be effective unless any potential propagation step has a low activation energy. Linear termination steps are likely to involve association of the chain centre with an exceptionally stable free radical or reaction with a molecule to give one or more inactive radicals, and so to have low activation energies. Thus the rates of linear termination can only be kept low by removing potential inhibitors. In gas-phase chain reactions the surface of the reaction vessel often behaves like an unremovable inhibitor, the effect of which can only be reduced by using large vessels or by hindering diffusion to the walls by adding inert gases. The walls are less important with chain reactions in solution because diffusion to the wall is very slow, and termination in solution is usually by mutual recombination. The rate of the propagation step is favoured relative to the mutual termination step because the reactant concentration is so high relative to the chain-centre concentration.

Thus the essential feature of a spontaneous chain reaction is a fast propagating step. This requires a low activation energy and, in terms of collision-theory concepts, a probability factor near unity. The low activation energy is achieved in exothermic radical–molecule reactions, but it cannot be achieved if the reaction is appreciably endothermic.

The profound effect of this can be strikingly demonstrated by comparing the potential chain reactions in hydrogen and chlorine and in hydrogen and iodine. The thermal reaction in hydrogen and chlorine at 300–400 °C is certainly a chain reaction. In hydrogen and iodine, at similar temperatures, it is equally certainly a non-chain reaction though it may be two-step (*see* p. 18). This is rather surprising because it requires less energy to dissociate iodine than to dissociate chlorine molecules. If a potential chain reaction were initiated by dissociation of the parent halogen, and terminated by recombination of the atoms, the reaction schemes and rate laws would be as follows (ΔH given in kJ mol^{-1}):

		ΔH		ΔH
Initiation	(1) $Cl_2 + M \rightarrow 2Cl + M$	243	(1′) $I_2 + M \rightarrow 2I + M$	147
Propagation	(2) $Cl + H_2 \rightarrow HCl + H$	−4	(2′) $I + H_2 \rightarrow HI + H$	134
	(3) $H + Cl_2 \rightarrow HCl + H$	−184	(3′) $H + I_2 \rightarrow HI + I$	−138
Termination	(6) $Cl + Cl + M \rightarrow Cl_2 + M$	0	(6′) $I + I + M \rightarrow I_2 + M$	0
Rate law	$2k_2[H_2]\sqrt{\dfrac{k_1[Cl_2]}{k_6[M]}}$		$2k_2'[H_2]\sqrt{\dfrac{k_1'[I_2]}{k_6'[M]}}$	

Thus the relative rates for equal concentrations of the reactants are

$$\frac{R_{HCl}}{R_{HI}} = \frac{\text{Rate of } (H_2 + Cl_2) \text{ chain reaction}}{\text{Rate of } (H_2 + I_2) \text{ chain reaction}} = \frac{k_2\sqrt{(k_1/k_6)}}{k_2'\sqrt{(k_1'/k_6')}}$$

$$= \frac{A_2}{A_2'}\sqrt{\frac{A_1A_6' \ \exp[-E_2 - (E_1 - E_6)/2]}{A_1'A_6 \ \exp[-E_2' - (E_1' - E_6')/2]}}$$

Now the Arrhenius A factors for 2 and 2′ are probably very similar, and likewise for 1 and 1′, 6 and 6′. The value of $(E_1 - E_6)$ is equal to the dissociation energy of the chlorine molecule, 243 kJ mol^{-1}, and $(E_1' - E_6')$ is the dissociation energy of the iodine molecule, 147 kJ mol^{-1}. Thus the relative rates are

$$\frac{R_{HCl}}{R_{HI}} = \exp(-E_2 + E_2' - 48\,000 \text{ J mol}^{-1})/RT$$

E_2 is known to be 23 kJ mol^{-1} and E_2' must be greater than the value of ΔH for step 2′, *i.e.* >134 kJ mol^{-1}. Hence at 500 K

$$\frac{R_{HCl}}{R_{HI}} > \exp(63\,000 \text{ J mol}^{-1}/RT) \sim 10^{6.5}$$

Simple energy considerations therefore suggest that a chain reaction is much more likely to occur between hydrogen and chlorine than between hydrogen and iodine. The iodine atoms are readily formed, but cannot react rapidly in the endothermic step, 2′:

$$I + H_2 \longrightarrow HI + H \quad (\Delta H = 134 \text{ kJ mol}^{-1})$$

and so they recombine by reaction 6′.

The equivalent step in the (hydrogen + bromine) reaction is less endothermic, as the bond in HBr is stronger than in HI. Hydrogen and bromine combine by a chain reaction where the chain length is fairly short. This reaction has been extensively studied and it illustrates very clearly how information about elementary steps in an established mechanism can be obtained by combining rate and equilibrium data.

The experimental rate law for the reaction

$$H_2 + Br_2 \longrightarrow 2HBr$$

when initiated photochemically is fairly simple if initial rates are examined. The initial rates are proportional to the concentration of hydrogen and to the square root of the intensity of light absorbed. The square-root law suggests that mutual termination of chains predominates (case *b*, p. 69). We are thus led to a scheme very like that for case *b*. However, later in each run the rate falls off more rapidly than predicted by the initial rate law, and if hydrogen bromide is added at the start the initial rate is decreased. As the equilibrium lies well over to the right at low temperatures the retardation is not due to reversibility of the overall reaction. However, just as with the hydrolysis of diphenylmethyl chloride, it does indicate a reversible stage in the mechanism.

The thermal reaction behaves very similarly. The initial rate is proportional to the hydrogen concentration, and to the square root of the bromine concentration; the rate falls off in each run from the prediction of the initial rate law, and also the initial rates are retarded by adding hydrogen bromide.

More detailed investigations show that the experimental rate laws are

Photochemical

$$\frac{d[HBr]}{dt} = \frac{k(\theta)^{1/2}[H_2][Br_2]}{[Br_2] + k'[HBr]}$$

Thermal

$$\frac{d[HBr]}{dt} = \frac{k''[H_2][Br_2]^{3/2}}{[Br_2] + k'[HBr]}$$

The form of the denominator suggests that Br_2 and HBr compete for some chain centre (just as H_2O and Cl^- compete for $(C_6H_5)_2CH^+$ in the hydrolysis of diphenylmethyl chloride, p. 50). These centres are likely to be H atoms, and the only possible reaction between H and HBr is the reverse of the initial propagating step $Br + H_2 \rightarrow HBr + H$. Thus the scheme for the thermal reaction becomes:

(1)	*initiation*	$Br_2 \xrightarrow{M} 2Br$	k_1
(2)	*first propagation*	$Br + H_2 \longrightarrow HBr + H$	k_2
(3)	*second propagation*	$H + Br_2 \longrightarrow HBr + Br$	k_3
(4)	*reversal*	$H + HBr \longrightarrow H_2 + Br$	k_4
(5)	*termination*	$Br + Br \xrightarrow{M} Br_2$	k_5

The rate of formation of hydrogen bromide is obtained by first setting up the differential equations for the rate of formation of HBr (formed in 2 and 3, removed in 4), bromine atoms (formed in 1, 3 and 4, removed in 2 and 5) and hydrogen atoms (formed in 2, removed in 3 and 4). The principle of stationary states is now applied to the equations, setting $d[Br]/dt = d[H]/dt = 0$, and this gives expressions for the stationary-state concentrations of the free atoms that can be substituted into the rate equation for the formation of product. The resulting expression is

$$\frac{-d[HBr]}{dt} = \frac{2k_3k_2[H_2][Br_2]\,(k_1[Br_2]/k_5)^{1/2}}{k_3[Br_2] + k_4[HBr]}$$

$$= \frac{2k_2(k_1/k_5)^{1/2}[H_2][Br_2]^{3/2}}{[Br_2] + (k_4/k_3)[HBr]}.$$

This equation derived from the proposed mechanism can be compared with the experimental rate law. It is clearly of the correct form, and the experimental values of k'' and k' are related to the rate constants of the proposed steps: $k' = k_4/k_3$ and $k'' = 2k_2(k_1/k_5)^{1/2}$. Now k_1/k_5 is equal to the equilibrium constant of the dissociation $Br_2 \rightleftharpoons 2Br$. This can be calculated, as explained on p. 2, and hence k_2 can be determined. A stationary-state treatment of the photochemical reaction and comparison with experiment shows that $k = 2k_2/(k_5[M])^{1/2}$ and so k_5 can be evaluated (strictly $k_{5,M}[M] = k_{5,H_2}[H_2] + k_{5,Br_2}[Br_2] + k_{5,HBr}[HBr]$ and each separate $k_{5,M}$ can be evaluated). In addition k_4/k_2 is the equilibrium constant of the reaction

$$H + Br_2 \rightleftharpoons HBr + Br$$

and as this can be calculated k_4 can be determined. From the value of $k' = k_4/k_3$ and k_4, k_3 can be determined.

It is then necessary to test the assumptions made in applying the principle of stationary states, namely that the concentrations of bromine atoms are low, so that their rates of change are negligible. This test is satisfactory. Finally, estimates have been made of the values of k_1, by dissociating bromine, and of k_5, by studying the recombination of bromine atoms (*see* pp. 4 and 7), to provide satisfactory tests of the values derived from the chain reaction mechanism.

5. Résumé of Aims and Principles

In discussing elementary and complex reactions many comparisons were made between the experimental rate laws and rate laws based on an assumed mechanism. While it is easy to apply the principles that have been discussed to a specified mechanism, it is much more difficult to decide whether a particular reaction proceeds by one mechanism or another. Experience has shown that a variety of experimental evidence must be accumulated about a reaction before its mechanism can be identified with any confidence. Of prime importance is the measurement of the rate and its quantitative dependence on any factors that are found to affect it. Usually the investigator has some likely mechanism and rate law in mind, based upon comparison with the known behaviour of similar reactions. The design of experiments and consideration of mechanism go hand in hand throughout the investigation.

It is first necessary to devise a reliable method of following the reaction and measuring its rate. The dependence of rate upon the concentrations of reactants and products must be found, and the site of the reaction located as a homogeneous phase or at an interface. The effect of temperature must be investigated over as wide a range as is allowed by the maximum and minimum rates that can be measured. The effect of possible intermediates in the reaction, of possible impurities acting as catalysts or inhibitors, and of different environments must be found. Thus for gas reactions it is important to change the shape and size of the vessel and the material of the walls to discover whether any reaction step is occurring on the walls. For reactions in solution, it is important to change the solvent to find whether solvent molecules, or any species derived from them, take part in the reaction. In aqueous solution, for example, changes in ionic strength may cause changes in rate and point to ionic intermediates, and changes in pH may reveal catalysis by hydrogen or hydroxyl ions. By using reactants with isotopically labelled atoms it is possible to gain information about the actual bonds which are broken in the reaction and so help to identify the mechanism. The thermochemistry of individual reaction steps of a mechanism impose certain restrictions on the activation energy, as explained on p. 28, and the accumulation of thermodynamic data about reactants, products and intermediates is of great importance in kinetic studies.

It must be emphasized, however, that the selection of a satisfactory mechanism depends upon scientific judgement as well as upon scientific experiment. There are several reactions—in fact, several classes of reactions—where the choice between possible mechanisms is in dispute among the various investigators. Moreover, it is now evident that many reactions, for long assumed simple, may not have simple unique mechanisms, but react by one route or another according to the temperature, the extent of completion of the reaction or even the absolute concentration of the reactants.

Elementary reactions

Elementary reactions proceed in a single stage through a transition complex to products.

On the *molecular scale* these reactions have the following characteristics:

(i) a small number of bonds are broken and formed. Suitable reactions include: bond scission, whether homolytic as in $Br_2 \rightarrow 2Br$, or heterolytic as in $(CH_3)_3CCl \rightarrow (CH_3)_3C^+ + Cl^-$; bond formation, as in the reverse of bond-scission reactions; the exchange of a simple group, as in $H + D_2 \rightarrow HD + D$, or of an ion, as in $CH_3COCH_3 + B \rightarrow CH_3COCH_2 + BH^+$, or electron, as in $Fe^{*3+} + Fe^{2+} \rightarrow Fe^{*2+} + Fe^{3+}$ where the star indicates an isotopically identified species; the simultaneous exchange of two such groups from one molecule to two others, as in $ON + O_2 + NO \rightarrow 2NO_2$;

(ii) the requisite number of molecules, ions or radicals meet simultaneously to form a transition complex; it is most important that any proposed complex should be geometrically and sterically possible;

(iii) in most elementary processes it is necessary for the molecules to share a minimum amount of energy, ϵ, in order to form the transition complex, and the fraction of collisions that are successful is proportional to $\exp(-\epsilon/kT)$;

(iv) if ϵ is zero or very small, the frequency of collisions will determine the rate. In these circumstances it is possible for diffusion of solute species through the solvent, or of a gaseous species through a mixture of gases, to control the rate of reaction.

On the *macroscopic scale* the experimental diagnostics for elementary reactions in gases or ideal solutions (*see* also p. 56f.) are:

(i) the rate law must be simple and be that predicted from the chemical equation for the reaction by setting the rate proportional to the concentration of each and every reactant;

(ii) the rate law must be obeyed over wide changes in the total and the relative concentrations of any reactants;

(iii) the rate law must be obeyed over a wide range of temperature;

(iv) the rate constant, k, usually varies with temperature according to the Arrhenius equation, $k = Ae^{-E/RT}$, where A is independent of the temperature or varies only slowly with the temperature in comparison with the associated changes in the exponential term;

(v) if the reaction involves the association of two or more species to form the transition complex, the factor A may depend upon an inverse power of the temperature, with the result that if E is very low the rate constant may decrease as the temperature increases;

(vi) when elementary reactions are reversible, the velocity constants, k_f and k_b, of the forward and back reactions fit the equation $k_f/k_b = K_c$, where K_c is the equilibrium constant of the reaction. Also $(E_f - E_b)$ equals the change in internal energy, $\Delta U^{\ominus}$, of the reaction.

Complex reactions

Complex reactions proceed by a number of linked elementary reactions in consecutive or concurrent steps. One very common pattern is

$$\text{Reactants} \underset{2}{\overset{1}{\rightleftharpoons}} \text{intermediates}$$

$$\text{Intermediates} \xrightarrow{3} \text{products}$$

The intermediates may be relatively unstable molecules or ions, very reactive free radicals or atoms, or unstable complexes like those between enzyme and substrate or between gases and solid surfaces. Many important reactions seem to follow this pattern, including many organic reactions, many charge-transfer reactions between ions, and most homogeneously or heterogeneously catalysed reactions. The equations can rarely be solved exactly. Computer methods have proved extremely powerful; in their absence, a simplifying assumption first used by Bodenstein has proved very useful. It is the principle of *stationary state* or *steady state* of the concentration of any reactive intermediate. The principle asserts that during most of the reaction the concentration of such intermediates can be considered as constant. Thus each of the terms d[intermediate]/dt can be neglected in comparison with other terms in the differential equations. This usually leads to expressions for the stationary-state concentrations of the intermediates, and hence to an expression for the rate of the whole reaction, as exemplified in in Chapters 3 and 4.

Another very important pattern gives chain reactions. In the absence of branching, a typical chain mechanism is

Initiation step: Reactant $\longrightarrow$ chain centre

Propagation step: Chain centre + reactant $\longrightarrow$ product + chain centre

Termination step: Chain centre + chain centre $\longrightarrow$ unreactive species
Chain centre + inhibitor $\longrightarrow$ unreactive species

The rate equation for any particular set of these steps is derived by applying the principle of stationary states to the chain centres. For a successful chain reaction the propagation step must be fast, as the initiation step is usually slow. In the presence of powerful inhibitors, chain reactions show induction periods during which little reaction occurs. The rate increases when the inhibitor is destroyed. A rate which increases from the initial value is also shown when degenerate branching occurs by the formation of centres from some fairly stable intermediate or product.

On the *macroscopic scale* the experimental diagnostics (not all of which may be applicable to any one reaction) of complex mechanisms are:

(i) a rate law different from that predicted by the rules for elementary reactions;

(ii) if the rate law is simpler than expected the slowest step occurs early in the sequence of steps. The rate law then indicates the composition of the activated complex in the rate-determining step (*see, e.g.*, p. 49);

(iii) a rate law that changes as the reaction proceeds or is different if the ratio of reactant concentrations is changed or if the temperature varies;

(iv) a rate law that is the sum of several different terms, each of the form $k\Pi$[reactants], indicates a number of parallel routes. All but one of these routes must be complex. Typical parallel routes are homogeneous and heterogeneous reactions; catalysed and uncatalysed reactions; catalysis by different species, such as general acids or bases; elementary molecular reactions occurring together with free-radical complex mechanisms;

(v) the presence of concentration terms in the denominator of the rate equation. These may be reactant concentrations, which usually indicate the saturation (or potential saturation) of a limiting pathway such as active sites on an enzyme or a surface; or product concentrations, which show either the presence of a reversible reaction at some stage before the rate-determining step, or competition with the reactant for an intermediate species or for a reaction site; or an inhibitor concentration, in a chain reaction; or an inhibitor or 'poison' concentration in an enzyme or in a surface-catalysed reaction;

(vi) if the rate of reaction is altered by changes in the surface/volume ratio of the vessel the reaction may be wholly or partly heterogeneous, or be a chain reaction with initiation or termination on the walls;

(vii) the suppression of the rate by traces of inhibitors usually indicates a chain reaction, unless the reaction is catalysed and the inhibitor is blocking sites on the catalyst;

(viii) a rate which increases as the reaction proceeds indicates a complex mechanism. The increase may be due to destruction of an inhibitor, formation of a catalyst, formation of chain centres from a product molecule or direct chain-branching;

(ix) an activation energy which changes as the temperature changes. If the mechanism consists of consecutive steps the rate may be controlled by the step with higher activation energy at the low temperatures, but by the step with lower activation energy at higher temperatures. If the mechanism consists of parallel paths, that with lowest activation energy may predominate at low temperatures, and that with highest activation energy at higher temperatures;

(x) a rate constant that decreases as the temperature increases may be due to a reversible association reaction to form an intermediate, followed by slow breakdown of the intermediate. In a few complex reactions (*e.g.* enzyme-catalysed reactions or some thermal oxidations) the rate constant first increases as the temperature is raised, only to decrease at higher temperatures. This is often due to the destruction of an intermediate species by some side-reaction.

Suggestions for further reading

Introductory reading

E. S. King, *How chemical reactions occur.* New York: Benjamin, 1963.

M. A. Atherton and J. K. Lawrence. *An experimental introduction to reaction kinetics.* London: Longman, 1970.

Reference Books

A. A. Frost and R. G. Pearson, *Kinetics and mechanisms.* New York: Wiley, 1958. Second edition, 1961.

I. Amdur and G. G. Hammes, *Chemical kinetics: principles and selected topics.* New York: McGraw-Hill, 1966.

D. Bunker, *Theory of elementary gas reaction rates.* Oxford: Pergamon, 1966.

E. F. Caldin, *Fast reactions in solution.* Oxford: Blackwell, 1964.

S. F. Benson, *Thermochemical kinetics.* New York: Wiley, 1968.

E. S. Gould, *Mechanisms and structure in organic chemistry.* New York: Holt-Rinehart and Winston, 1959.

F. S. Dainton, *Chain reactions.* London: Methuen, 1955. Second edition, 1966.

K. J. Laidler, *Chemical kinetics.* New York: McGraw-Hill, 1950.

S. L. Freiss and A. Weissberger (eds), *Investigations of rates and mechanisms*, (vol. VIII of *Techniques of organic chemistry*). London: Interscience, 1953. Second edition, 1961.